DEDICATION

For all whom the Creator will say: *"Well done, good and faithful servant! You have been faithful with a few things; I will put you in charge of many things. Come and share your master's happiness!" Matthew 25:23*

ACKNOWLEDGMENTS

Let's start by acknowledging our Creator for whom we would have no life. Every life is a gift from God. I re-discovered Jesus Christ when I was 49 years old. My search for the reasons his Holy Spirit came over me has been ongoing for all this time. When I came to know God's truths, I did not think I deserved his Grace. Then I learned the true meaning of "grace". God left us the greatest book ever written for mankind, his Holy Bible. This is the most amazing book as it has the ability to speak to us each time it is read, bringing new meaning that applies to our lives.

There have so been many ministries and authors who have taught me so much. For all those with pure hearts and love of God, even though we may sometimes see things differently, I acknowledge and praise. Creation Ministries International is doing fantastic work. Answers in Genesis and Ken Ham have done amazing work that resulted in the building of the "Ark" in Kentucky. Other mentions are D. James Kennedy, Jon Paulien, Doug Batchelor, C.S. Lewis, R.C. Sproul, Hal Lindsey, et al. I could fill the page. God knows the work of those I have not acknowledged! Praise God for all these good and faithful servants.

Creation Cosmology
101 – 202

By: Charles N. Slayton

TABLE OF CONTENTS

Introduction

An informed person will admit that current scientific claims about the beginnings of the universe and life are filled with innuendo, speculation and some falsehoods especially when it comes to allowing the public to accept so many unproven theories as truths. An informed reader will also know that cosmologist can disagree as much as any two politicians. The *Creation Event,* implies a Supreme Being created the Universe. The event will be explained using information available in today's physics and cosmology. I am not reinventing the wheel! Let's start with an appeal for an equal footing between claims made in this work and claims made by most of today's cosmologist. It seems science and engineering fields are flooded with folks who have accepted naturalistic beginnings as truths but disagree vastly on the details. In truth, cosmologist know their science is not factual but seem perfectly okay if the public thinks so. There is much to be gained by close examination of the postulates brought forth here. In most every case, there will be sufficient scientific and mathematical backup to support reexamination of contemporary theories.

Since this work is about *Creation Cosmology,* we are going to put God's point of view on the table. Chapter One will start by claiming the Holy Bible is the truthful account of Creation. According to the Bible, God himself told Moses the sequence of his Creation and Moses recorded it in what is now the Pentateuch which is also the Book of Genesis. Bible chronology indicates the creation event happened about six-thousand years ago. In the Seventeenth Century, Bishop Usher [34] developed a timeline based upon Biblical chronology. His timeline is considered a fairly accurate estimate for the time of Creation.

If all of the Universe was created in place about six-thousand years ago, then gravity and light from each star and galaxy are racing out through space at the speed of light. So far that light has only reached out in distance about six-thousand light-years (6 KLY) from each star. We can then postulate that the vastness of space must still be void with no gravity influence. Of course this account of Creation presents a dilemma for how we can see light from galaxies and stars more distant than six-thousand light-years. We clearly see objects in space that are estimated to be millions and billions of light years away. Solving this paradox is the purpose of this work.

A Paradox of Light Speed

So what is this paradox and what does this mean? By definition, a paradox is a seemingly absurd or self-contradictory statement or proposition that when investigated or explained may prove to be well founded or true. In my opinion, the last frontier of *Creation* needing further explanation is the field of cosmology, specifically relating to how light would have traversed the Universe in times much shorter than what we know as the speed of light. This work considers the relationships between light and gravity and may offer reasonable explanations. We will see that gravity waves and light are tied together as they have the same speed. A young Universe would require that light somehow traveled across the universe much faster than the speed of light would allow it today.

This presentation on cosmology is done with the reader being presented information in a casual yet science-based way to both dispel the old and support new ideas about our beginnings. It's success will be measured at a later date and time after much review, analysis, criticism and perhaps new revelations have taken place. For example, if cosmologist are unable to discover necessary dark matter as the fabric holding the Universe together, it may be necessary to throw the baby out with the bath water and look for completely new explanations. The forthcoming postulates, science and reasoning are presented as the door is wide open for better explanations of our beginnings.

Virtually every field of *naturalism* has been successfully challenged cutting deep flaws in those theories. By showing that the earth is not as old as claimed, we open two very important arguments; first, life would not have had time to evolve on a young earth and second, we could not exclude design by our Creator. Not only could life not have evolved on a young earth, neither could it have evolved anywhere in a young Universe.

Within all areas of science, discoveries have been made that bring into question the validity of *naturalism* [1] and *evolution*. Any reasonable mind digging into it just a little and applying basic understandings will see numerous flaws. Naturalism is the state of life today without a creator having influenced any of it but on the contrary, everywhere we look, we see evidence of design. Just a few years ago, we considered life to be much less complex than what has been revealed by today's science research [2]. A not forgotten common claim was that life began in a puddle of some kind of protein soup. Any reasonable mind looking into it will see life is far too complex to seriously consider naturalism. That soup was missing an enormous amount of information required for the very simplest of cells. No one on earth can assemble and create life even in its simplest form.

We are going to discuss cosmology as it is known today and review some of it's facts and theories in order to prepare ourselves. This book is a bit like military boot camp. Coming in, we are going to tear down whatever you thought you knew about cosmology, then rebuild you with principles based upon reasoning and evidence of our Creator God. We will be considering a series of postulates based upon reasonable, speculative and inductive reasoning that may open doors for new understandings of time, cosmology and our beginnings.

Just as in present theories of cosmology, an informed reader may find error within this work. This work is approached from years of observing science and cosmology with a discerning eye in search of truth if it be found. Regardless of any criticism to come, I strongly believe there is merit is this work that will cause minds to think, rethink and explore new avenues. Knock and the door will open [3].

The Order of the Report

An attempt has been made to simplify this work enough for the informed layperson to follow. Simple drawings and diagrams are used frequently to help in understanding. The subject of cosmology is quite complex in light of today's advancements. Importantly, we have had only a fraction of time to determine our beginnings based upon the observational Universe regardless of how old you may speculate it's age to be. Following the first chapter where our created beginnings are explained, we will build the case for looking elsewhere to find answers as there is so much misinformation out there about the beginnings of the Universe. Following that, we will begin to examine the actual physics as it applies to both contemporary cosmology and *Creation Cosmology*. Then later, we will give our Creator more credit and get into speculative ideas of how God did some of the things he did.

Theories and Faith

Once a theory is proven, it becomes a fact. Faith can be said to apply equally across all theories about our beginnings. Contemporary cosmologist and evolutionist's claim the earth and the universe are billions of years old. Einstein claims his theory of special relativity must hold throughout the Universe. Most cosmologist claim the Universe began as a single mass that expanded throughout the Universe but now they need to find invisible things such as matter and energy to hold it together. Considering how far reaching many of these claims are, let's see if contemporary discoveries have opened doors to allow for God's work.

Why should Biblical history and Scriptures be considered ?

Critics often claim we simply cannot commingle Creation testimony found in Bible Scriptures with science. If this is the case, then reasonable minded people of faith in God should profoundly conclude that *"science will never reach the truth of our beginnings"*. By starting out building a wall around a frame of reference that excludes the role of our Creator, one cannot logically reach the truth provided there really is a creator. So in fairness, we should at least conduct our affairs of research from both perspectives to see if either hold water. Logically, science can never prove there is no creator as we cannot prove a negative. Because of the lack of complete evidence of our beginnings, all avenues must be faith based.

How much has been injected into today's science without evidence? Much of that will be discussed throughout this work including Dark Matter [4]. Wouldn't Dark Matter be considered to be as supernatural as Creation? As we will further discuss, Dark Matter and the Big Bang [5] have no advantage over the Biblical Creation account [6]. Reasonably ask yourself, what has been readily rejected by science? The answer is the Holy Word of God our Creator.

So how should scripture-based Believers unravel two-hundred years of science-based attacks on the Creator? In order to do so, we must necessarily expose the level of faith built into evolution and naturalism just as faith is built into our Creator's testimony. Scripture accounts and the Big Bang cannot both be true at the same time.

Postulates

Postulates are used in this work as a starting place for logic and reasoning with new ideas about space and time. Postulates will always be accompanied by materials providing a basis or foundation for their reasoning. If we postulate that a big bang must have happened, then that is our starting place of reasoning. If we place the starting date at a different time, then that time will be the basis of our reasoning.

Illustrations and Drawings

You will find illustrations and drawings throughout. A few data tables are also included. Others are borrowed and credit is given to their originators. In most cases, the drawings will help explain some of the more complex thoughts and ideas.

Math

For those who may feel insecure with math, there is some use of it in this work but it is not essential to follow the work as all will be explained. To simplify much of the math used in this work which is not a lot, just remember that anything multiplied by zero equals zero and anything divided by zero violates math rules and is undefined or not possible.

When is math just math and science just science? Math is our universal tool for explaining relationships between all sorts of things especially in the fields of science. We measure everything using math; for example our networth, our physical conditioning in units of abilities, the distance we can drive on a tank of gas and the caloric value of the foods we eat. Whenever we define something new, we measure its success or its worth by assigning some degree of value using math. The most common unit of measure is the value of money with respect to goods and services we want or need. We determine the value of going to the Moon or to Mars by evaluating the distances and the cost/benefit of getting there. We do the same when we buy a car or plan a family vacation.

Math can easily be manipulated to present things in their most favorable light. This is why we need wisdom and discernment to analyze and assign value to most everything in our lives so we are not deceived by people with agendas that may limit or distort the facts we need. All materials being covered here can be explained using very basic physics and algebra. Let's not complicate this. If you are not so mathematically inclined, do not be alarmed, as all will be explained in hopeful terms and diagrams the reader will understand without the math.

Circular Reasoning

Circular reasoning is a logical fallacy in which the reasoner begins with what they are trying to end with. The components of a circular argument are often logically valid because if the premises are true, the conclusion must be true. Wikipedia

Circular reasoning is very common in science and mentioned here as it is often used in justifying cosmology. A common example is when an unproven theory is used to prove another theory. You may see it in this work as well, so watch for that, too.

Chapter 1

The Universe according to the Creator

What would a "Created Universe" look like? It can look no different from what is seen out there today. We have been watching the Universe for a few hundred years now and there is no way to tell how it began. We have only one Universe and this is it. In order for readers and skeptics to have a clear understanding of the Universe as seen through Scriptures, this chapter is going to take the Genesis account of the Bible which is the literal Judeo-Christian account as the truthful word of our Creator God in Heaven.

The difficult pill to swallow in following the logic of this work is the premise of open space with absolutely no gravity where time does not exist. Based upon the time of Creation of the Universe, there are vast regions where gravity has not yet reached. This concept is completely alien to most cosmologist.

Everything presented here will be analyzed in later chapters. The best way to start building *Creation Cosmology* is to first understand it by having it explained from a Biblical/Scriptural perspective. Skeptics be patient as we need to introduce you to this universal view of Creation. Set aside your criticism for now until the entire case is presented and explained. We will see new ideas that challenge many areas of cosmology and that should peak any reader's interest.

According to Biblical chronology, the Universe and all that is in it, was created about six-thousand years ago and people have lived on this earth for that same amount of time. Since Judeo-Christians believe the word of God is the truth, then Noah's flood had a major impact on the appearance of the earth seen today. As fascinating as that study is, our focus will be looking out into space.

Around six-thousand years ago, by the will of God, the Universe came into existence along with everything in it within six twenty-four hour days. On the first day, He created the heavens and the earth. In the days that followed, God created an amazing variety of life and everything we see on earth today. This creation was very good until mankind turned away from the love of our Creator. Our Creator gave us free-will and mankind through free-will, separated himself from the love and fellowship God intended for us. Except for saving one faithful family, God chose to destroy mankind causing a great flood. Evidence of that world-wide flood is everywhere. The subject of this work requires us looking up into the heavens. That's where we will go for confirmation of what is to follow.

Before creation took place, the space needed for God's creation was vast darkness containing nothing, no mass, no gravity, no energy and no light. It could be called non-dimensional space. Obviously that space from our perspective was enormous as it was needed to contain what was to come. If there was anything in that non-dimensional space prior to creation, it was placed there by the Creator as part of His plan.

Figure 1: A star like our sun. Courtesy of (nasa/esa)

Imagine a star as in *Figure 1*. Our sun is a star, imagine it at creation. At the instant God brought it into existence, a massive *gravity field* shot out radially and continues to travel outward. At its outward expanding boundaries, its strength has significantly weakened and may be very difficult to detect as far out as six-thousand light-years which is the distance its gravity would have reached by now.

At the very instant of creation, the sun's nuclear fires also lit. *Energy* from its nuclear fusion engine independently expanded outward at the speed of light. On the surface of the earth, we can see the sun's energy in the form of light from about eight minutes ago. That energy is put to good use heating the earth. Energy that is not put to use somewhere in the solar system is gone from us. We can speculate that the Universe contains massive amounts of energy that is not put to use. We will be looking further into this.

The expanding gravity field of the sun by itself can attract nothing until it reaches another object of mass having its own gravity field. The sun's force of gravity attraction is ready to be applied to any body of mass within reach but until it does, it can do nothing. After eight minutes following Creation, the sun's light and gravity reached Earth and the Moon. It had already captured Mecury and Venus and is now racing out toward Mars then beyond.

For purposes of this discussion, let's call the expanding gravity fields traveling at the speed of light, "Bubbles". We will also call them gravity shells or gravity fields which are expanding.

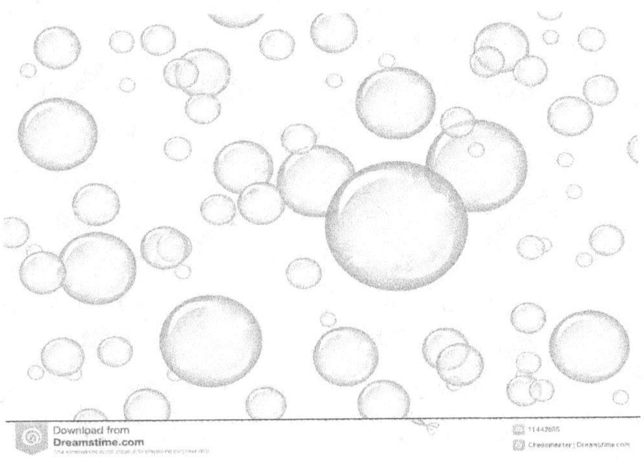

Figure 2 Bubbles of expanding gravity fields (dreamstime.com)

Gravity Shells are expanding Bubbles

We can define the bubble as the radius of the effective limit of gravity around a star. Energy inside of the bubble will react with gravity to produce light. We will see that energy in the form of light passing outside of the gravity bubble will not produce light. Light passing through a gravity bubble will obey the rules of physics relating to refraction and diffraction. We will explain more on this later.

Figure 2 represents each object of mass in the created Universe enclosed a bubble of gravity that is expanding outward at the speed of gravity waves which is the same as the speed of light. Any two objects less than six-thousand light-years away from each other have already begun experiencing a pull, a mutual attraction toward each other. Any objects of mass that are farther out remain unaffected by mutual gravity attraction.

Since we can see star-light from much farther apparent distances than 6,000 light-years (6 KLY), we need to ask ourselves how did light from those stars get here so fast? The first thing that comes to mind is since gravity has not reached out that far yet, light must have some way of moving much faster in the vast voids of space outside the expanding 6,000 year-old bubbles. The answers we seek seems to lay claim to light and energy having some independence from gravity.

Light and gravity waves both travel at the speed of light [5]. It is a fact that when we turn on a flashlight, its light travels at the speed of light. Then we remember that light speeds change in different densities of materials such as in air, water, glass, etc. However, every test or measure of the speed of light done today is within the effects of gravity. So perhaps light can travel much faster when it is outside the effects of gravity or the bubbles. This unknown about light becomes a focus of discussion with the conditions scriptures have set in this chapter.

Before we get into the details, let's look at what the observation results would be for an observer on the created earth at different times. In summary, this creation event allowed massless energy to permeate the universe ahead of expanding gravity fields traveling at the speed of light. What did Adam see on day six? When Adam looked up on the first clear day and night, he could see the sun and the moon and the entire universe pretty much as we see it today. How could he see the universe? If our earth had only been in existence for six days after the universe first appeared, the earth and Sun's gravity field would have only extended six light-days out. Adam's sky would be like a giant 3-D screen with all the energies of stars and galaxies appearing perfectly in the sky. If we continue the logic of this explanation, then Noah, Abraham, Jesus, John F. Kennedy and you, the reader would all have seen pretty much the same sky seen by Adam and Eve. The explanations for this are coming.

Galaxies and Gravity

At the moment, we are left with little to go on other than light which is a part of energy, somehow advances itself above the speed of gravity out there in space where gravity does not yet exist. By *faith*, we can draw this postulate:

Postulate A: Energy and light can only be constrained by the forces and speed of gravity when in its presence.

Perhaps at the moment, this sounds over-simplified but is it? This postulate cannot be argued since cosmologist believe the Universe is very old so this postulate obeys or complies fully to their thinking since they assume gravity is everywhere. We will be getting into much detail and discussion in future chapters. How big of a role will *faith* be playing in this work? How about *supernatural* events? Contemporary cosmology is yet to explain the Big Bang. Therefore, the Big Bang is a supernatural event and requires faith to accept as theory because it is not a fact, but this is jumping ahead into future chapters. There are other meaningful ways to look at our beginnings and we will.

God's primary focus on his Creation of the Universe was on this world. He created it for us and he created us in his image. He created this world to his unique design because he wanted it perfect for fellowship with his Creation. We are significant. He set his laws in motion and undertook the immensely complex task of making us living breathing souls along with all other life on this earth. Once accomplished, he realized he had done well and was ready for fellowship with his Creation. He created the physical universe in full splendor all around us. He set us at the center of it all. This world is and was set apart and remains that way until God decides otherwise.

Postulate B: God would have known the influence of sin upon mankind who would rebel and seek to circumvent His authority. God would have left evidence for the enquiring mind to find and justify his work. So we seek in earnest. The very survival of the Holy Scriptures over the millenia is a miracle itself.

Know this for certain, science will never be able to prove how our origins came about; we can only speculate. This chapter describes briefly what people of faith in the Creator are told by Holy Scriptures and what we should believe and teach. We all admit that faith is applied to religion. Now we must see if cosmology itself is a religion. From here on, we will be examining what contemporary cosmologist believe about the truth of *their* scriptures. Of course we cannot lump all skeptics into the same category, but we certainly know they have given absolutely no credit for God's Creation. This is unfortunate.

Conclusion

Holy Scriptures testify the Creator created the Universe about six-thousand years ago. Gravity has only extended out to six-thousand light-years. God's design of the Universe and his laws of physics under certain conditions allow energy and light to travel much faster than the recognized speed of light.

Chapter 2

How should we approach the Science of Cosmology?

People like to believe in all sorts of things like ghosts and UFOs. Of course there is no proof of either. A few years ago I attended a show in Las Vegas starring famous illusionist, Lance Burton.[8] It was very entertaining, he made himself vanish then re-appear in another location, then a Corvette vanished and appeared at another spot. Now as impressive and entertaining as this show was, I knew these things were illusions and not real? Obviously it came from my own sense of reason and logic based upon information I have acquired throughout my life up to the time of the show. Is it possible that someone would see this show and conclude that these things magically happened? Most likely yes, because you use your own ability to reason to conclude that a person with limited knowledge who was exposed to or encountered such things might accept it as fact.

What if a person was taught and believed a car can disappear and re-appear somewhere else. Would that mean that the car really did disappear? No, what it actually means is false information has been spread to that person and presented as fact. Of course, there is no way to prove what really happened to the car while observing from an audience.

Consider that *we* are the audience in the contemporary field of cosmology.

How does this magic relate to science and cosmology today? If I am taught something that sounds reasonable but is incorrect and I rely on that information to reason with the world I live in, I am seriously challenged since I have been subjected to false teachings leading to false conclusions. Not only that, I have excluded valuable information resources either by my own ignorance or simply by not looking for other reasonable explanations. This makes me victim of shaky and perhaps dishonest information. The solution is to question everything that seems unreasonable.

Cosmology

Cosmology is the study of the universe as a whole. In doing so, we consider where it came from, what's happening within it and what is going to happen to it in the future. Cosmology is also defined as the science of the origin and development of the universe. Modern cosmology [7] is dominated by the Big Bang Theory, which brings together observational astronomy, particle physics and speculation. Contemporary cosmology as well as *naturalism* would have us believe there was no creator involved. This should present a huge dilemma for people of faith in the Creator.

Einstein

What did Einstein not consider? It is becoming apparent that energy, in the form of waves may only exist in fields of gravity. All of Einstein's equations must include gravitational forces to be coherent.

Remembering *Postulate A: Energy and light can only be constrained by the forces of gravity when in its presence.* We will be getting into how to deal with that in a young Universe.

Postulates, Axioms and Theories

A postulate or axiom according to Wikipedia is a starting place or point for reasoning. When we construct a postulate, we suggest or assume the existence, fact, or truth of something as a basis for reasoning, for discussion, and belief. It is not vastly different from a theory which is defined as a supposition or a system of ideas intended to explain something. In any case, theories and postulates are dependent upon a system of ideas that most always become the basis for some degree of "faith" in what is sought at the outcome. Throughout this work, postulates will be presented. We will evaluate them for their validity using facts with both inductive and deductive reasoning to support or at least question conflicting postulates and theories. It is important to remember that once a theory or postulate is proven, it becomes a fact. Cosmology is built upon theories and postulates that are not facts, they are assumptions.

The task before us is to further develop thoughts and ideas about how energy behaves and particularly light which is part and parcel of the energy spectrum. A key player in this development is understanding the roll of gravity since everything we know and have learned historically, has been within this very small laboratory we call the solar system which is completely influenced by gravity. Astronauts in orbit around the earth are subject to gravity, even the Voyager crafts launched decades ago are still subject to gravity. We will methodically see if we can draw some new ideas or at least open our minds to further investigation.

When is thinking "outside the box" appropriate?

We are about to take a journey into an examination that requires thinking outside the box. In the process, we cannot limit our thinking or restrict ourselves to what seems to be well established facts about physics. We begin by proposing some *what ifs* and considering them in light of what is known. We carefully examine what is known within the sphere of evidence, and looking outside the box to see if we can develop new ideas that may be very difficult to prove with current methods for testing and analysis. Does this sound out of the ordinary? History is filled with examples of breakthroughs that resulted by thinking in profound ways and putting those beliefs to the test. There is no blame for such thinking.

A very common but not the only cosmological *belief* today is that there must be something called Dark Matter [4]. Cosmologist claim the order of the Universe does not make sense without it. By throwing in Dark Matter, it helps make sense of what some cosmologist have grounded themselves into believing. They must *believe* that dark matter exists or their theories fall apart. They necessarily cling to widely accepted but unproven theories on the basis they will make sense when they discover Dark Matter. We will look at how science came up with the over-simplistic idea of the Big Bang in the first place.

Let's define what is science and what is not science

In the name of Science! More false and improvable claimed facts about the history of earth have been espoused by scientist. They ignore exculpatory evidence in order to conform to predetermined beliefs that are not provable.

The real problem we face as Creationist is today's science cannot accept a young universe that did not take billions of years to bring itself into existence. The issue then becomes the placement of time, *t equals zero* of our beginnings. If *(t=0)* was billions of years ago, we must ourselves, "create a beginning" that led us up to at least six-thousand years ago. Little has changed in space for the past six-thousand years. If *(t=0)* is six-thousand years ago, we must look at space much differently. This view affects motions of galaxies, light travel, gravity fields (Bubbles), red shifts, time dilation, expansion of the Universe and on and on. These changes resulting in the relocation of *(t=0)* to about six-thousand years ago is the purpose of this work.

When the scientific community promotes a theory as truth without solid scientific evidence, that is not science. There are popular faces we see in the media and on documentaries practicing the reinforcement syndrome of the unproven and that narrows our critical thinking. Today's scientists and physicist especially on shows produced for television like to lay out a few basic provable theories then jump to radical unsupported ideas about the past, present and future. These pronouncements and claims are a reflection of their own *faith* or lack of *faith* in another sense.

Consensus Science [9] is generally the mutual acceptance of theories within the community of the like minded. Cosmologist are judge and jury to any other opposing ideas or exculpatory evidence that may be contrary to their faith-based theories. The simple proof of *consensus science* as being okay with theories presented as fact is found by looking into our school and college text books. Consensus science is what we are all taught and led to believe but it does not make it truthful. It is a matter of *faith*.

Creation Science [10] which is simply putting God into all fields of science, has produced an overwhelming amount of information that is contrary (exculpatory) to today's generally accepted theories, especially in dealing with *naturalism*. *Consensus Science* has built a wall around itself blocking any matters of faith while at the same time employing its own type of faith to pursue sometimes, nothing more than voodoo science. You cannot debate it, because it is "walled-in" asserting there is no comparison between naturalistic science and the "flawed" six-thousand year Scriptural view of *Creation Science*. Hitler's propagandist Joseph Goebbels supposedly claimed, that if you tell a lie big enough and keep repeating it, people will eventually come to believe it. Mark Twain is reputed to have said: *"Never let the truth get in the way of a good story."* The work done here is to search for truth regardless of where it takes us.

How critical is it that we give Creation Science an opportunity to be heard? Most people in the world claim to be Christians. Cosmology is declaring the Bible cannot be truthful. The Bible tells us to test ourselves, and the truth will be revealed; if it is not truth, that also will be revealed. *(2 Corinthians 13:5) Examine yourselves to see whether you are in the faith; test yourselves. Do you not realize that Christ Jesus is in you--unless, of course, you fail the test?)* This is the work to be done. If we can show postulates that actually support the possibility that light traveled much more quickly across the universe, then the Big Bang loses its magic and so does the age of the universe. More importantly, it draws us closer to the Creator. There is nothing in the Universe more important than that.

Creation Science

Creation Science is the interpretation of scientific knowledge in accord with belief in the literal truth of the Bible, especially regarding the creation of matter, life, and mankind in six days. In today's scientific community and often in the religious community, this belief is quickly dismissed. By dismissing and accepting naturalism, the consequences are enormous. For example, they dismiss the questions of why are we here and what happens to us when we die. The God of the Bible has promised to those who believe His Word, an eternal life in a new world without sin. Sounds pretty good to me!

Ethics in Science

So you may ask what does ethics have to do with gravity and light since so many think it is "settled science"? We explain using an analogy of self-serving agendas.

Let's say you vacation at a foreign place of interest and hire a tour guide with a reputation of leading informative tours. You have placed yourself in a position of dependence upon him. Then ask yourself, "is he best serving the tourist or is he serving his own agenda", which is typically his livelyhood. The tour guide is analogous to contemporary scientist in charge of interpreting everything they want us to know about their established field. They are going to show and explain what they want you to know in a way to encourage you to support their work. We place our trust in them for presenting us information. Like the tour guide, if they are doing their job, they leave you with a positive impression because that sustains their funding. We can safely claim that self-serving agendas are inherently built into humanity and is therefore a very large part of science.

Are we claiming that all science is false because it is riddled with self-serving agendas? Not so, but consider, are scientist morally any different than politicians or used car salesmen? We'd like to think so and we will look at that a little later. However, if you seek to know the truth regardless of where the consequences may lead, then you have applied high standards of ethics and principles to your work. Truthful and honest science, just like politics, should require holding to a higher standard of accountability. Scientists claim that supernatural forces cannot play a role in science, but at the same time claim the universe began from nothing and yet must be filled with undiscovered dark matter. Are they any different in applications of *faith*? Not at all.

Perhaps contrary to the confidence placed in theories most of us are presented with, modern science is riddled with circumstantial evidence, innuendo, corollaries, and theories about things we do not know and may never know. As I read and researched for this work, I was surprised by the amount of speculation physicist and astronomers inject into expanding their knowledge about our world and universe. Overall, I have great respect for scientist. They are imaginative people and often inject their own faith into their logic like claiming or postulating the universe must be full of life while it has not been proven. This work is no different except we are basing our faith on the word of God.

The purpose of this work is to present how new and innovative theories can lead to other conclusions, other starting points and other time-frames. Bible scriptures tell us a

Universal Creator played a role in what we are and what we see today. The skeptical minded should at least give consideration to this work. What harm is there in looking at new theories and postulates and perhaps using the same level of faith employed in the Big Bang theory.

Creation Science believes that debates on our origins are not influenced by man alone but came about due to both good and evil spiritual influences in our lives since the beginning of man. Perhaps then, all of us should give greater consideration to the possibility of spiritual influences participating in our work. The question is, which spirits do we want influencing us? If cosmologist are truly claiming that no creator or spiritual influences were involved, then ethically, they should also deny their own faith in unprovable facts about our beginnings. Some actually do. *(Psalms 19:1-2) The heavens declare the glory of God; the skies proclaim the work of his hands. Day after day they pour forth speech; night after night they reveal knowledge.*

What harm has come from believing contemporary science taught in our schools and colleges today? If we can establish solid ground for having faith in Bible scriptures about our beginnings, our Creator will smile upon us. The problem is that to believe otherwise is to discount Bible scriptures as untruths. Jesus himself referred to Moses stating that *"if you believe in Moses, you believe in me; for he wrote of me"* *(John 5:46).* Jesus did not deny anything said by Moses including the Flood. Noah's Flood is a certainty. The goal of this work is to put God in the highest esteem of morality and truth. If God did create, then we must work to prove it as far as possible. Perhaps we should realize the dead-end believing in the Big Bang since it now requires non-existant or undiscovered dark matter to hold the universe together. We will be looking at this.

One would think that by now with the use of modern super computers, we should be able to simulate galaxy and solar system formation. This can be done by programming but in order for galaxies and solar systems to form, we need fudge factors, or "external" nudges, as galaxies and solar systems would not form and spiral from randomly distributed materials left over from a big bang. What we are seeing today appears to have required a programmer influencing galaxy formations to produce exactly what we see. Without some order of information or coercion, data bits in a computer simulation would always remain random and disorganized. How is it that there are millions of galaxies all acting in order by some information that was necessary in the first place to organize them? Logic says there must have been some organizing external forces for the universe to appear as we see it today. Cosmology and Newtonian physics have not explained it.

Evolution as it applies to light speed research

Lets start here by discussing *evolution*. The word itself is very broad-termed as it is used in fields of geology, biology and the development of processes including thought. In the evolution of thought processes relating to science, sometimes we become so entrenched on a theory that we are unable to think outside of the box. We have put our ideas and mindsets about how things seem to work into a box or cage and we limit our notions about our existence and how things must work.

This is fine in some fields of science and engineering, but becomes elusive and evasive when used in the cosmology. When we make that mistake we handicap ourselves. When Creation Science proposes to express concern over a well entrenched theory, we are subjected to *consensus science* opposition by our brothers and sisters within a particular field.

A well established mistake scientist make because of their entrenchment in a particular belief is to jump to radical conclusions advancing their beliefs to the extent they abandon science and journey purely into faith without connection to the facts. An example of this is "since there is life on earth, there must be life throughout the universe". A similar radical jump would be to believe that atoms are actually tiny solar systems containing even smaller life or our universe is but small pieces to something much larger. These are radical yet similar in they both require faith to believe in something beyond the current limits of our knowledge and science. Many applications of thought in the fields of cosmology are wild speculation and innuendo but deemed necessary to sustain some of the present improbable theories.

Naturalistic science is always changing dates. The testimony of God does not change dates. So what is truth? Is it the man-made sciences of evolution and naturalism or is it *"In the beginning God created?"* The goal of this writing is to show how science may be reconciled. How will these differences play into the inevitable future of mankind and to each and every one of us on our appointed day? We will summarize this later.

Conclusion

If we start with the premise "there is no god", it cannot be a logical arguement because you cannot prove a negative and it has excluded evidence of God. The evidence is highly reliable testimony from credible witness recorded in Holy Scriptures. It seems the sciences involving naturalism including cosmology have been led down an exclusive path of reasoning which has caused an effect we call *Consensus Science*. To protect their rules and influence, they have created exclusive clubs. Only the like-minded are allowed into their club. This club of consensus science must accept naturalism as the golden rule and that rule is simple, it rejects the Creator and His testimony. Ironically, the club has not rejected the application of *faith* applied in its own work.

Evolutionist start with the premise *"the world looks old, so it must be old and because light red-shifts, all science must comply with an old world and universe"*. This is unqualified acceptance of one type of faith and complete rejection of another. If the Universe is truly young, then all sciences based upon an old universe are simply circular reasoning based upon a false premise. It is based upon a few theories that cannot be classified as facts leading to false conclusions. Lets start with a fresh look.

Chapter 3

Problems with Today's Science

Today's Theories Critiqued

In principle, what is different between Creation Cosmology and the theories of the Big Bang? This Report challenges science to discover how energy and light actually do react outside the presence of gravity. If the Universe is young, then gravity cannot exist everywhere. This presentation has built a case for the reality of it.

So how has naturalistic science dealt with creation science when it conflicts with their widely accepted beliefs that also employs faith? The frame work of this question is to show that when creation scientist present evidence that conflicts with the very heart of naturalistic science, there is a right way and a wrong way to deal with the discrepancies. First, naturalistic scientists typically ignore creation science. They do not want to draw attention to it. After all, in the opinion of most, creation scientist have produced stumbling blocks that if given enough press would seriously impact their science.

Unfortunately, naturalistic science is the basis of *atheistic ideology* brought to light very early during our education of the sciences and has been passed down as the scientific gospel. The purity of science is flawed due to its lack of tolerance for opposing matters of faith.

Naturalistic scientist claim they cannot consider anything such as Creation or a Designer as it is not science in itself. In essence, they try to close the argument before it happens. Isn't this in itself an act of extreme prejudice? Where is the "tolerance" for transparency of facts that dispute their master plan of naturalism? This approach allows them to ignore their guilt of employing very strong matters of faith in theoretical areas such as the Big Bang and Dark Matter. The fundamental question has always been regarding the frame of reference with which we begin our postulations. Naturalist will argue falsely that to inject a *designer* into science is to place constraints on their abilities to think and reason about science. That is just not so.

There are so many holes in naturalistic beliefs about our beginnings and the time it took for our present world to appear as it does today. An excellent resource that covers all fields of science is a book called *"Evolution's Achilles' Heels"*. [13] Rather than labor here to repeat so many revealing facts that are widely known and established, I refer the reader to materials presented in the "Reference" section at the end of this work. There are many references that make excellent cases for opening the minds of today's scientist to the reality of weakness and implausibility of naturalistic science when excluding the Creator.

Rejection of Creation Science

Today's scientist ignore many facts in evidence proven by Creation Scientist. It is difficult to present this book without feeling some animosity toward the damage being done to society by atheism and agnosticism. When injected into the minds of the general public, it deliberately draws us further and further away from the knowledge of our Universal Creator. Why are so many of today's leading public figures in the sciences so agnostic in their thinking? They have built a wall of separation between science and God as Creator. They make their own rules yet they violate those rules every day. They are guilty of consensus science by supporting the basic flawed concept that all things must be naturalistic meaning their rules do not permit any minds or greater intelligence than what their own minds can conceive. Yet is this true? I think so, as they speculate on all sorts of wild imaginary things as dark matter, big bangs, etc. These things when seen as truths, are their own "gods of the science world".

Should we consider biblical scriptures of past events as evidence or hearsay? Consider that God spoke directly to Moses telling him the order of the created the Universe. That is not hearsay, Moses is a fact witness. The written accounts in Scriptures remain as evidence and fact to Moses' testimony. If you argue that it is hearsay, then you must necessarily argue naturalism without proof is also hearsay. We have not thrown out all evidence of cosmology or all of biological naturalism as much of it speaks to facts. So lets not throw out scriptures until we have put them to the test. Let's use scriptures as a frame of reference for a new starting place as we are trying to accurately solve the "mystery of creation" of the Earth, the Universe and space and time.

Christ is the Light of the world. God is light, and Christ is the image of the invisible God. One sun enlightens the whole world; so does one Christ, and there needs no more. -Matthew Henry Commentary, Concise

The Scientific Method

Operational science is verifiable by repeated experiments. It is what we can see and measure with a degree of accuracy and expected outcome. In cosmology, what we see versus what we can measure is different, therefore we are limited in many ways particularly in the sciences of evolution and cosmology. To expand our knowledge, we apply inductive reasoning since we can only use what is available to us as measurable data. Many scientist build their theories on circumstantial and speculative information that does not meet the standard of the Scientific Method and in some cases are to far stretching to be considered reasonable. However, every idea has to start somewhere and develop over time by building supporting facts.

Is there dishonesty in today's science?

Should we readily accept everything science tells us today? Maybe it's best we do not and for lots of reasons. We will get into looking at reasons, so let's start by taking a look courtesy of *"The Journal of Clinical Investigations"* and *"Psychology Today"*.

How do we study dishonesty?

The Journal of Clinical Investigations* 11*

https://www.ncbi.nlm.nih.gov/pmc/articles/PMC4639975/

*(**abbreviated content used here**)Throughout the years, we have gathered a vast amount of evidence for this deeply human condition. We study dishonesty with various conflict-of-interest tasks (1). One such task is the matrix task, in which participants are presented with a test sheet with 20 matrices, with each matrix consisting of 12 numbers. Participants' task is to find and circle in each matrix two numbers that add up exactly to 10 (e.g., 1.53 and 8.47). For each correctly solved matrix, we pay $0.50. After five minutes, participants count the number of correctly solved matrixes on their test sheet and write down their performance on a collection slip. In our baseline control condition, participants submit both the test sheet and the collection slip. The experimenter then checks each participant's performance and pays them accordingly. In our treatment condition, we instruct participants to shred their test sheet and only submit the collection slip to the experimenter. In this latter condition, participants face a conflict of interest at the end of the task: they know that they can either be honest or overstate their performance to earn more money. Several versions of this experiment, with thousands of participants in various countries, consistently find that, despite random assignment to these conditions, participants "solve" more matrixes in the conflict-of-interest condition — evidence for dishonesty. Interestingly, however, despite theoretically being able to claim having solved all 20 matrixes and getting away with it, people rarely cheat by the maximum amount. On average, they cheat by only 2 to 3 matrixes. **Finally, these observations are never driven by a few bad apples that cheat by the maximum possible amount while others are honest. Instead, we find that almost everyone cheats but only by a limited amount.**

> *Psychology Today [12]*
>
> *https://www.psychologytoday.com/us/blog/moral-landscapes/201807/honesty-and-truthfulness-in-science*
>
> **What other qualities go along with honesty in science?**
>
> *Posted Jul 22, 2018*
>
> *THE PRIMARY AUTHOR OF THIS POST IS DR. TIM REILLY.*
>
> **(ABBREVIATED CONTENT USED HERE)**
>
> *Concerns about scientific ethics are receiving more attention, especially amidst high profile examples of methodological favoritism (designing studies of alcohol use in consultation with alcohol producing companies) and even data fabrication, such as a study falsely claiming new uses for stem cell technology.* **Instances like these of dishonesty create obstacles for other researchers whose work hinges on the reliability of published research.**
>
> *Discussions in science ethics tend to focus more on DIShonesty than on a positive description of honesty.* **In hour-long conversations with scientists about what makes for good science and the qualities of a good scientist, data fabrication has been called, "the cardinal sin of science".**

Scientist should recognize self-deception by knowing when "things" just do not add up. This is "discernment". That is a gift from God inherent in all of us except perhaps an unfortunate few.

We have the best economic system available as it allows us personal freedom to pursue free thought. As a capitalist society, we have benefited greatly. Most scientists are driven by the self-sustaining need for money and most of today's money for science comes either directly or indirectly from taxpayers. Sometimes, in order to get money, scientists may exacerbate the potential benefit of their work to attract attention needed in order to convince lawmakers and taxpayers the value of their science. Just like in any other field of free enterprise, competition is very stiff. It may be necessary to spend huge amounts of preparation time to magnify the importance of their science in order to obtain the money needed to sustain their work. [14]

The Standard model for the Universe

The *standard model* [15] in the minds of Big Bang cosmologist, assumes the Big Bang to be true. The error is when they establish parameters they need to achieve this. Any evidence that does not fit into their walled-in ideas may be conveniently thrown out. This approach has so far led cosmologist nowhere and has excluded other possible considerations about our beginnings. The Big Bang is soley based upon *faith* in supernatural beginnings and as a result, this model assumes that we came from primordial soup!

Naturalistic Formation of the Universe

The first stumbling block evolutionist in the field of cosmology encounter is claiming all material we see in the universe came from a single point that exploded with a Big Bang. They are claiming the Universe came from what they call a singularity which is defined as a point at which nothing, a zero-value function, takes on an finite or enormous value and fills the Universe. They are very careful not to claim the universe was created from nothing, somehow it began as a point of zero or some unknown mass and spread out to the size of the universe with us included. That's supernatural!

No science on earth today can explain this supernatural event or where the material of the universe came from. The Big Bang in fact, is a supernatural event, if it occurred. The only clue of the Big Bang event is from observing *red shifts* in star light of galaxies far away. The Doppler Effect is present in both sound waves and in light waves. When an object produces either light or sound while the object is moving away from us, the sound/light waves stretch becoming longer. This causes light color to shift toward the red spectrum. This red shift gives us the appearance that stars and galaxies are moving away from us. So if that motion is reversed, it leads them to believe all matter in the universe must have come from the very same point at an earlier time. That in itself is amazing as it would appear that we could possibly be at the very center of the Universe.

If we think about reversing the Big Bang for a moment and claiming the Universe collapses on itself, we can say with scientific certainty that when matter is condensed, especially all of the matter in the universe, it would go into some state of nuclear fusion as stars do. The Universe would temporarily become a super massive star and in time become a black hole, as black holes in theory are super dense masses of spent matter. Returning to a singularity would require something going back to nothing and that defies known physics. So, in the beginning, did a super massive dead black hole explode producing light elements once again capable of re-forming active galaxies and stars? The answer to this question cannot explain our beginnings but works well for recycling. We cannot truthfully know or explain our beginnings. Maybe if we throw in a little magic or create some unknown phenomena or even some form of catalyst we could produce naturalistic creation on a computer but it would not be proven science. This is justification to look elsewhere seeking our origins. The Big Bang cannot be explained by physics as it is not repeatable.

When we look back in time using current theories of light travel, what we see appears to be a lot of homogeneity throughout the universe; a nearby galaxy appears very similar to a galaxy estimated to be out to ten billion light-years or more away from us. Lets think about how we can best explain this. Consider we start with two identical galaxies yet they are alleged to be billions of years difference in age. Because of red-shifts, if one of the galaxies has a much greater red-shift, then it must be much farther away and since the light took billions of years to reach us, the galaxies cannot be the same age. However, in a young Universe, they can be the same age and the light got here much sooner. Then there must be other explanations for the causes of red-shifts.

Creationist Tas Walker stated "In fact, it's downright cruel to teach children that they evolved from animals, that there is no design or purpose to their lives, and that the universe is coldly indifferent to their existence".

The eternal benefit of believing in our Creator God and taking seriously his account of our very existence should drive any reasonable person to give it the consideration it deserves. As a believer in the Creator, I cannot imagine my life without him. There would be no purpose. We owe it to our Creator to know him and understand his purposes for our lives.

Stars

Nuclear physics has benefitted us significantly in understanding how stars sustain nuclear fusion reactions. Our focus in this work will be on the actual energy produced by stars specifically in the form of light and the star's gravitational forces over great distances.

Black Holes

Black holes are quite a phenomena when we consider what they do to light. Scientist claim that black holes have gravity so strong that even light cannot escape. If this is the case, then light must have mass or some weight that gravity acts upon to literally slow it to a stop. Are they saying that light speed can change and even stop? They do not say how energy, when created, can accelerate from a star to the speed of light in the first place. How can light go from zero to the speed of light or from there back to zero? So does light have mass or doesn't it? Einstein says objects containing mass (material matter) cannot accelerate to the speed of light. We will be looking into that.

The answers to light speed we are seeking in this work lies in its relationship with the *speed of gravity* [16]. Contemporary physics tells us that gravity waves have a speed and so does light and they happen to be the same speed. We briefly questioned how light is influenced by gravity around a black hole, but can light exceed the speed of gravity. If gravity effectiveness approaches zero, what happens then? We will get back to that. By considering that light actually rides on gravity waves explains even greater universal importance of gravity.

Cosmology and the Size of the Universe

Looking deep into the Universe is interesting and may produce some useful understandings about it. The study of cosmology is truly fascinating. Let's ask the following question: Is it necessary to look so deeply into space to determine the validity of *Postulate A': Energy and light can only be constrained by the forces of gravity when in its presence.* We are actually seeking answers to the question of what happens to the properties and speed of light once it escapes gravity.

Postulate C: Energy and light escaping gravity fields cannot conform to Newtonian Physics.

This postulate has a mathematical basis as both Newton's and Einstein's relativity fail when no gravity is present. This subject is coming up in following chapters.

The trouble with Photos of Stars and Galaxies

The photos we commonly see from deep space are greatly enhanced by astronomers expanding the visual light spectrum normally seen with our own eyes. Light on both sides of the visible spectrum are colorized into the visible spectrum in photos giving us a much wider range of viewing. Since viewing these wider ranges of the energy spectrum requires special types of cameras, this leaves amateur astronomers at a disadvantage. The focus of this work is the energy that is produced within the visible light spectrum. However, we will see that all energy responds to changes in gravity field strength just as visible light.

How do we measure light speed and distances in space?

This of course brings up the "what we know" about the measurement of light and how many different tests we have to do this. There are ways we measure energy and wavelengths that are not visible to us such as radio waves and wavelengths on either side of the visible spectrum. Some are radio waves are bounced off the moon and satellites, laser beams reflected off of satellites, deep space satellites with synchronized earth based clocks. None of these tests are able to determine what happens to light outside of the effects of gravity. In later chapters, we will look at conditions in micro and zero gravity to see what may happen when gravity fields become so weak they can no longer carry light and energy.

Looking at distance measurements, we can reasonably measure the distance to nearby stars by observations from the earth as we orbit around the sun giving us a form of triangulation to calculate distance. This is the *parallax method* [17]. This method in itself includes many variables and complicates the task but does not make it impossible. Just as we can measure the speed of light, we can also measure the distance of nearby objects by measuring the time it took for radio signals to return (s=d/t). We can do this for local objects in space, but all of these are done in our neighborhood in the presence of gravity produced by our sun, moon, earth and other planets in our solar system.

We will need to define some terms for distances used in understanding how large the Universe appears to be. One *light-year* [18] (LY) is the distance light can travel at the speed of light in one year. 1 LY = 5.9 trillion miles. The size of the Universe is claimed to be 13.6 billion light-years, that's (5.9 x 13.6 = 8) followed by 22 zeros in miles!

Very accurate telescopes using *paralax* can resolve distances pretty accurately in space for stars out to about 326 light years. 326 light years is a miniscule fraction of size of the claimed Universe. Beyond that, there is no precise way to measure distances. Other less accurate methods for more distant stars compare their light properties and intensities to similar nearby stars using spectrometers. They also measure distances based upon estimated age and motion determined by a star's red shift. These less accurate methods are estimates built upon an accumulation of other estimated data (circular reasoning). We do know for certain that the Universe is very large and other galaxies are very far away.

Red Shifts

As briefly discussed, Red Shifts [19] are seen as wave lengths of light stretching due to relative motions between a star or galaxy and the observer. An object moving away from us causes light waves to shift to the longer or red range of the light spectrum. Blue light indicates an object is moving toward us.

Our interest in red-shifts is to understand what could happen to light traveling over great distances in micro or very low to zero gravity. This is the light "outside" of the gravity bubbles previously discussed. We will be looking at what happens to light as it passes near objects in deep-space in order to determine the effects on wave lengths of light. This could change what we know about red-shifts.

Speed of Light Tested. What do we know and not know about light speed?

20

All tests of the *speed of light*   (C) have been within our local region of this vast universe. While speed test have shown slightly different results, its exact value has been set at 299,792,458 meters per second (approximately 300,000 km/s (186,000 mi/s)). We cannot argue as radio waves operate at the same speed as far out as Voyager space crafts. We are going to look at some very different conditions as we proceed and see if we can determine what changes may occur to the speed of light.

The speed of light for fish analogy

Is the speed of light universally constant? The speed of light tested passing through water is much less than in air or a vacuum [21]. What if we evolved in the ocean and were intelligent fish and we measured the speed of light in our environment, we would come up with a lower speed. So how is this relevant? Just as replacing water with air gives us a huge difference in speed, what would happen if we were able to measure the speed of light in micro-gravity or zero gravity? Are we just like the fish in not knowing what really lies out there beyond the limits of our lab?

We will be looking into this. We must consider the places in the vastness of the Universe where gravity is either extremely weak or does not exist such as "outside" of the bubbles.

Big Bang

If the Big Bang were true, it means that at an instant in time, something decided to create itself from nothing. The material from nothing contained just the right amounts of hydrogen and heavier elements that would miraculously experience gravitational forces sufficient to condense and eventually form massive nuclear fusion reactions. That process would miraculously transform the trillions of tiny fused particles into other elements and that would transform into other elements and so on until beginning with nothing leaves us a habitable wonderfully designed world. A world occupied by intelligent human beings with an amazing universe in which to gaze. This is how some theoretical physicist claim both past and future universes would form and react. Perhaps there were billions and billions of big bangs earlier and to come after this one but only this one produced people or did all of them? Logically, this line of thought is so far out there, maybe it's best to consider a Creator.

Ironically when confronted with the theory of a big bang, most physicist have their own doubts, yet it is the most prevalent theory and it is the popular "headliner" for modern cosmology. It is the theory that is necessary for excluding God from the equation. It forms the basis for our godless beginnings as presented in the science of *naturalism*.

Rotation of the Universe

If there was a Big Bang, was it spinning at some hyper-velocity during it's time before the event? What is causing spiral galaxies to rotate? Could there be some universal "external" *Coriolis Effect* [22] causing galaxies to rotate? The Coriolis Effect is commonly observed by the rotation or vortex of water going down a drain. Do all galaxies rotate in exactly the same direction due to an unknown external force? What about galaxies tumbling or flipping like a dish flipping side over side? Of course it is not possible to answer any of these questions because no one saw what happened nor is there enough order seen in the universe to even make a prediction. The most common galaxies are spiral that have rotation but the source of the initial rotation is unknown.

With our present degree of knowledge, we simply do not know how galaxies would have formed with their given observable rotations, especially since there is no consistent alignment or pattern that might hint that some outside forces similar to a Coriolis effect initiated such rotation. Galaxies are randomly distributed and each spiral galaxy is rotating with its own structure and order although not all are perfectly formed spiral galaxies. A few galaxies are seen as colliding with each other, so there is no clear order to the distribution of galaxies in the Universe. It is impossible to scientifically conclude that materials distributed throughout the Universe formed and organized itself into what is seen today. There is no proof.

Galaxy Formation and Doom

Cosmologist are always going to try to prove the existence of the Universe and all of its contents in a naturalistic way. I understand that and I also understand that so far, all attempts have failed. Failure includes lack of evidence as applied using the scientific method. Once we depart from scientific methods of proof, we are left with theories and postulates and most of all, some type of faith. We know absolutely nothing about the original conditions or the contents of the early Universe. We know nothing about initial conditions of the states of matter, or the states of motion, nor do we know the actual time $(t = 0)$ which is the time the Universe came into existence. There is no proof that the Universe is billions of years old or just six-thousand years old. Other than red-shifts, there is no rational scientific explanation as the application of faith is necessary regardless of viewpoint.

Dark matter and Dark Energy

Why do we need dark matter? We need it to make current theories of cosmic gravitation work to hold the universe together. Dark matter is claimed to be a force or fabric that contains the mass and energy necessary for contemporary cosmology to make sense. So far dark matter has not been found.

The cosmology *standard model* theory now seems to demand that the universe is about 5% ordinary matter, which is observed through telescopes; 22.5% is dark matter, which is not observed; and the remaining 72.5% is mysterious dark energy. The need for the dark energy has been invoked by a need to explain the acceleration of distant galaxies away from us in all directions. This acceleration is solely based upon observable red-shifts. There is no hard evidence of any kind to support this additional needed matter or energy. There is no proof.

Prof. Stephen Hawking in his book "A Brief History of Time" said: 'This [Big Bang] picture of the universe ... is in agreement with all the observational evidence that we have today,' but admitted, 'Nevertheless, it leaves a number of important questions unanswered ...

The observational evidence Hawking spoke of is the apparent expanding of the universe evidenced only by light waves shifting to longer wavelengths. Without any scientific explanation to the origin of stars and galaxies, it follows there is no explanation for the motion, origin and structure of the universe. There is no proof.

"It's a huge mystery exactly how stars form,"says Dr. Richard Bond of the Canadian Institute for Theoretical Astrophysics.' This confirms what Creation Ministries International has said, "Stars could not have come from the 'Big Bang.'"

Exotic theories

Cosmological theories are becoming more radical than ever. In fact they are more like a house of cards built on speculation, innuendo and circular reasoning. Radical is the word for how 'far out' some of their models are. There are models that start before the Big Bang, where the universe supposedly arose from a fluctuation that may continually occur creating multiple universes. There is not a shred of experimental evidence for these theories and doubtful if ever. There is no proof.

A paper was posted to the Los Alamos pre-print archive (shown below) on 1 August 2002. The abstract below states that some assumptions of the inflation model lead to deep paradoxes. The last sentence makes a statement.

'Present cosmological evidence points to an inflationary beginning and an accelerated de Sitter end. Most cosmologists accept these assumptions, but there are still major unresolved debates concerning them. For example, there is no consensus about initial conditions. Neither string theory nor quantum gravity provide a consistent starting point for a discussion of the initial singularity or why the entropy of the initial state is so low. ... **Some unknown agent initially started the inflation high up on its potential, and the rest is history.**

Conclusion

We leave this chapter with the understanding that science in the field of cosmology is at best speculation about our beginnings. The postulates to follow deserve as much consideration as anything science has given us thus far. We need to discover more about this "unknown agent" mentioned above. Scriptures tell us who this agent is, he is the Creator.

The world looks forward to a dismal fate as the universe eventually dies in a 'big rip', 'big freeze' or a 'big crunch'. Without trust in the Creator and His Word the world has no hope. The only comfort is that the ultimate death of this universe is expected to be so far off in the distant future it can be ignored. But the Bible describes catastrophic world-changing events not so far off, only a matter of years, not billions or trillions. God has told us this in His Word. And though there are several different interpretations among Christians, the general consensus among those who take God's words as they were intended to be understood is that it is only a matter of years before major changes are expected on Earth and even in the universe (John Hartnett, Creation.com 30(3) 2016)

This dismal fate is not at all about *climate change*, it is about prophecies foretelling the limits of God's patience with those who have turned away from him and what he is going to do about it. When we come to know about God and his plan of *salvation*[38], we can come to understand that he has made promises for those who believe and accept his promises. For the others, he will punish for not believing. If you think this is foolishness, you are the problem and no good will come from you.

Hearing the truth can make you angry or it can open your eyes. It is God who will judge you or perhaps has done so already. Do not misunderstand me, I do not intend to imply that you must believe what this book is claiming. My goal is to provide you with a foothold on a better understanding of the role of our Creator God.

This chapter makes me wonder why our school text books are filled with this level of speculation about our beginnings. This faith in naturalism is not helpful in any way to society. In fact, it is driving otherwise reasonable minded people away from the only standard for morality. People have been taught for centuries about the moral authority of God. Sinful humanity has replaced that with its own and constantly changing values making it worse for society. Humanity is rapidly developing the attitude that *"no one else is going to tell us what to think or do!"*

Chapter 4

Opening the Door to New Ideas

Postulate D: Contemporary scientific theories lack sufficient evidence to prove claims made by naturalism, evolution or a big bang. This lack of evidence postulates the door is wide open to examine and test other possibilities regarding our beginnings and the existence of our Universe.

This postulate has been near the center of our discussion so far and has provided us a convenient point to start looking elsewhere. An appeal is again made here for the reader to allow equal footing by applying the same reasoning to this work as with contemporary cosmology.

When considering cosmology, why would we be so concerned about what seems to be well traveled and well proven closed-argument theories presented by elite groups of physicists and scientist? When accepted as fact by the public when it is not fact, it becomes an issue of ethics and may actually conceal the truth of our beginnings.

So how much damage has been done by science holding to faith in evolution and an old Universe? The damage is enormous! I cannot understate that fact! Any godless theory without solid evidence undermines its relationship with the Creator and that is damning. Scriptures present solid *eternal evidence* recorded by Moses who was chosen by the Creator from the time of his birth. This evidence is recorded in Scriptures for the benefit of all mankind.

Our very existence is the cause and effect of a designer much smarter than ourselves. In fact God is so intelligent that we would never know of his presence without the information left for us in his inspired written work. To ignore this record is to neglect his superior intelligence and replace it with speculation. We learn not only of his supernatural cause and effect creating the Universe, we learn of his purpose and promises. Briefly stated, his purpose is his desire to have a relationship with his Creation. His promise is the gift of eternal life. If you would like to know more about God's plans for sinful humanity, check out my book *Clearing Up Confusion about the Sequence of End Time Events.* [23]

Much more important to our lives is the question of what happens to us when we die. Some scientist may not realize the damage they are doing while others know but choose to ignore this most important question. By doing so, they are refusing to face the inevitable. They must wake up to the fact that they are rejecting the best source of information about the reasons for life and the very existence of our world and Universe. They are ignoring historical testimony from highly reliable witnesses who spoke prophetically not only about events before they happened, they spoke of events before the beginning of time and future events to come. They also spoke about his return to claim his own. Within each and every one of us is a spirit known by God. He knows us and he knew us before we were born. When our bodies die, there will be two resurrections, one for Believers and later one for sinners who will receive final judgment. He told this to us.

Many scientist are making a grievous error. They are in effect trying to reverse engineer the biblical account of creation into a naturalistic framework but they cannot do it without injecting their own supernatural values. An analogy would be alchemists ignorant of the properties of the elements trying to make gold but only successful in producing a bright shiny object. The bright shiny object is today's theories of evolution and cosmology; it's just too superficial yet it is held to with *deep faith* like it is a proven religion and it is not.

The skeptic will immediately say there is more evidence of an old universe and earth than the scriptural view of six-thousand or so years, but have they really taken an honest look? It is apparent they have not. There are vast amounts of science research supporting a young earth. Let's look at some resources that support the improbability of evolution as fact. This first post is by a non-religious person.

> **https://www.nature.com/articles/444265d** *Sir, In your News story "Polish scientists fight creationism" (Nature 443, 890–891; 2006 doi:10.1038/443890c), you incorrectly state that I have called for the "inclusion of creationism in Polish biology curricula"...* **I am critical of the theory of evolution as a scientist, with no religious connotation.** *It is the media that prefer to consider my comments as religiously inspired, rather than to report my stated position accurately.* Maciej *Giertych PHD Institute of Dendrology, Polish Academy of Sciences*

https://www.nytimes.com/1989/12/15/opinion/l-theory-of-evolution-has-never- been-proved-151289.html

Theory of Evolution Has Never Been Proved *DEC. 15, 1989*

December 15, 1989, Page 00042 The New York Times Archives

To the Editor:
In regard to "Evolution Theory's Foes Win Textbook Battle in California" (front page, Nov. 10): It is past time that those who purport general evolutionary theory to be fact be brought into the light. **Scientifically speaking, this theory does not qualify for classification as fact. It deals with history, which is not subject to investigation by experimentation.**

https://en.wikipedia.org/wiki/Evolution_as_fact_and_theory#Evolution_as_a_collection_of_theories_not_fact

Evolution as a collection of theories not fact*[edit]*

Evolutionary biologist Kirk J. Fitzhugh[39] writes that **scientists must be cautious to "carefully and correctly" describe the nature of scientific investigation at a time when evolutionary biology is under attack from creationists and proponents of intelligent design.** *Fitzhugh writes that while facts are states of being in nature, theories represent efforts to connect those states of being by causal relationships:*

*"***'Evolution' cannot be both a theory and a fact.** *Theories are concepts stating cause–effect relations. Regardless of one's certainty as to the utility of a theory to provide understanding,* **it would be epistemically incorrect to assert any theory as also being a fact,** *given that theories are not objects to be discerned by their state of being."*

*Fitzhugh recognizes that the "theory" versus "fact" debate is one of semantics. **He nevertheless contends that referring to evolution as a "fact" is technically incorrect** and distracts from the primary "goal of science, which is to continually acquire causal understanding through the critical evaluation of our theories and hypotheses." Fitzhugh concludes that the "certainty" of evolution "provides no basis for elevating any evolutionary theory or hypothesis to the level of fact."[40]*

ttps://creation.com/Cosmologists-Can-t-Agree-and-Are-Still-In-Doubt

(abbreviated report for purposes of highlighting cosmology problems)

Cosmologists Can't Agree and Are Still In Doubt!

Nearly 100 years since Einstein's theories were published, the origin and structure of the universe still eludes cosmologists

by John Hartnett

On 23 July 2002, NYTimes.com hosted an article entitled "In the Beginning ..." by Dennis Overbye. This was an attempt to put down any belief that science doesn't have the answers, i.e. it was a defence of scientism. The article pushes the point that even though, in the past, cosmologists may have been divided and lost on explanations of the origin, age and evolution of the universe, now this is not so.

Hartnett says:

"Agreement on fundamental cosmic numbers?

This is hardly the case. ....there are the issues of dark matter, the interpretation of peculiar red-shifts, even the interpretation of red-shifts themselves that are not agreed upon by cosmologists."

A reader should by now have sense to realize that cosmology is a very loose science with little proof. But still, this is what we are taught in our schools.

Conclusion

We have spent a lot of time establishing that naturalism is not entirely science but is a faith-based system of ideas that have been pressed upon society since the time of Darwin. Darwin's ideas were heavily influenced by atheism. As we move ahead, keep in mind we are using as much of today's science as possible but looking at it from a new, much more recent start date.

Chapter 5

Gravity and Light

At last, you have arrived where we delve into the differences of view points about how light is capable of crossing a young universe so quickly. Some important information was left for us giving us a basis to work from:

"My own hand laid the foundations of the earth, and my right hand spread out the heavens; when I summon them, they all stand up together."

I am sure many of you are not familiar with this statement. The purpose of this work is to acquaint you with this person by showing you the credibility of his work. You will see where this statement is found in scriptures toward the end of this work.

What is it in the Universe that causes light and energy to vibrate into wave forms? Is there some mysterious universal property *co-existing* that is undiscovered that contains all of the known frequencies that somehow overlays and manages energy? The below articles are a refreshers on properties of light which is a part of the energy spectrum.

> ***Light is formed by vibrations of electric fields that have a certain frequency.*** *Depending upon the frequency of the light the color of light is determined i.e. different frequency of light have different color..The velocity of light depends upon frequency of light as well as the wavelength of light, as there are different colors in light similarly there are different wavelengths in light. White light is composed of waves of different wavelengths, different wave-length of   light   creates   different   color.* ***https://www.tutorialspoint.com/ray_optics/dispersion_of_light.asp***

> ***All waves** are **caused** by some type of **vibration**. **Vibrations cause** a disturbance in the medium that becomes the source of the wave. Think about water **waves** formed when you throw a rock into a pond. The rock hitting the surface **causes** the water to **vibrate.***
>
> *Vibrations and Waves: Energy and Motion - Video & Lesson ..Search for:* ***https://study.com*** *› **academy** › **lesson** › **vibrations-and-waves-wavelength-am...***

Note that by this definition, all waves are caused by "some type" of vibration. Physics research has shown that gravity waves travel at the same speed as light.[16] This chapter is fundamentally important as it is based upon contemporary thought and discoveries in physics and cosmology about properties of light and gravity.

Gravity waves are nearly impossible to measure. Gravity typically appears to be in a steady state but it is not as other things are going on within gravity. Think of a gravity wave this way: at your particular location, gravity is like the surface of a very calm lake. On your side of the lake, you place a very sensitive instrument that can sense any movement in the water surface. A small rock is thrown in on the other side and a small ripple forms moving toward you. Eventually it reaches you and there is movement in the water in the form of a small wave or ripple. This change in the water surface is like the change in a stable gravity field. Also, by knowing the distance across the lake and time of occurrence, the speed of the ripple is calculated and in this example is found to be the same as the speed of light.

Gravity is normally very constant acting on objects on the surface of the earth. If you are standing on a scale and the scale momentarily changed as in a spike, that would be a gravity wave, but again, changes are typically so small they are almost impossible to measure. The point is that without a source causing a gravity wave and knowing its distance from us, we could not measure the speed of gravity. So normally we can only measure gravity as a force of attraction between two objects of mass. For example the gravity force of earth attracting a man. In principle, gravity is not measured in waves but measured as a force of the mutual attraction between two bodies. Let's review the current thinking about the properties of gravity. We commonly hear that gravity forces extend out to infinity but this cannot be true. Gravity fields or waves can only extend out at the speed of light from the time of their creation regardless of that age. That's still not infinity. As gravity fields extend outward, the field strength diminishes inversely proportional to the square of the distance. Gravity forces become extremely weak as distances increase. Our premise for this work is that gravity to date, has only reached out about 6000 light-years.

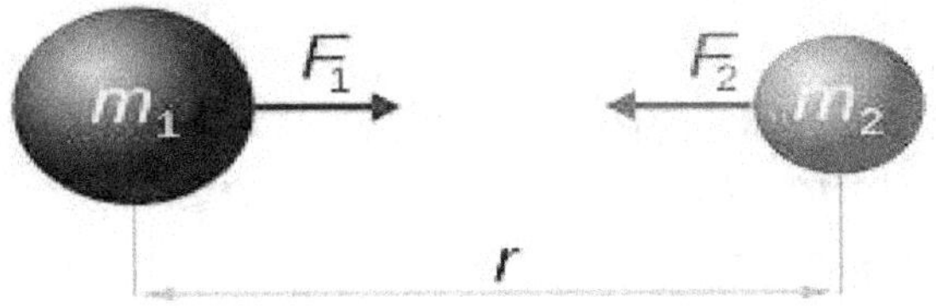

$$F_1 = F_2 = G\,\frac{m_1 \times m_2}{r^2}$$

Newton's law of universal gravitation states that every particle attracts every other particle in the universe with a force which is directly proportional to the product of their masses and inversely proportional to the square of the distance between their centers.

This formula shows at great distances the effects of universal gravity acting between two bodies will approach but never reach zero. This equation is valid and effective to determine forces between any two bodies of mass (m). Consider if we replace either one of the masses (m) with energy which has no mass, the entire equation is reduced to zero.

In the above formula, also note that (r) is the distance between two bodies in space. Note that in the equation (r) is squared, so its value becomes very large as distances increase. For example if the distance is actually 100 Light-years, since it is squared, the value in the equation becomes 10,000 light-years. The denominator value is this equation increases very rapidly as distances increase and that quickly reduces the forces of attraction. We can say that gravity strength decreases exponentially as the distance between the objects increase.

The above definition only reminds us that very large objects of mass have very strong gravity while small objects are minimally effected. If light has no mass, then what is its relationship with gravity? We are concerning ourselves with what happens to the motion of light when Newton's laws or Einstein's Relativity outside of gravity do not affect it.

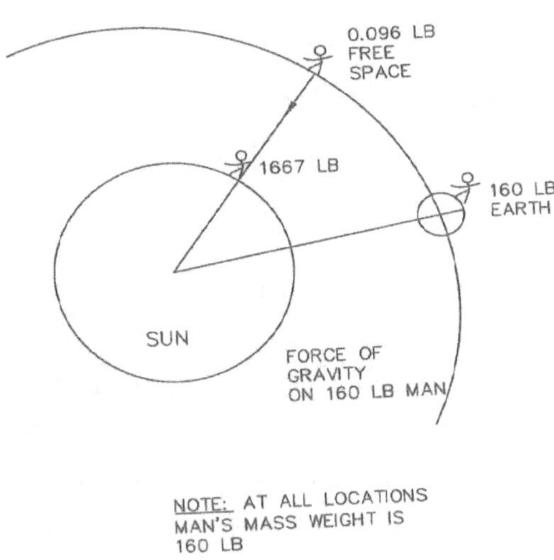

Figure 3: The weight of a 160 pound man in space

Referring to *Figure 3,* if a 160 lb man on earth was placed in a static position at the same distance from the sun as the earth but far around or away from earth's orbit position, his weight due to the attraction of the sun would be 0.096 lbs. Of course, if he could stand on the sun or at the radius of the sun's surface, he would weigh about 1,667 lbs.

So in reality, a 160 lb man (earth weight) floating in gravity-free space would actually have mass since he creates his own gravity. If we did not have *inertial mass* [28] , which is basically our "weight", then nothing would attract us. To attempt to calculate his mass and weight, we would need a secondary object of known mass by which we would then be able to measure the common attraction between the man and the object of mass. Sounding crazy now? A bathroom scale is used to compare our weight (inertial mass) with earth's gravity attraction. More will be explained.

For objects with small mass, at great distances away from planets, stars and galaxies, their mutual attractions become infinitesimally small. If the gravity attraction becomes so small as to have no influence on an object of mass, how does this affect light and energy? We will need to determine if there is an effective limit where gravity forces no longer excite or "vibrate" energy into producing light.

We will also be taking a look at how energy (and light) may not be able to organize or synchronize itself into wave forms but somehow with gravity it does. Energy (light) may be completely dependent upon gravity to organize itself into familiar wave-forms. If pure energy is massless, then what causes it to react in a way that makes it visible and capable of doing work? We will explore this question a little later.

Figure 4 below shows the relationship between possible changes in the speed of light affected by the forces of gravity. It looks at the range of gravity field strengths from zero gravity up to its maximum possible value. As gravity strength approaches its maximum, as in a black hole, light is claimed to not escape, so it must not accelerate. As gravity approaches zero, it is possible that light is no longer constrained by the speed of gravity. If light is stopped and cannot accelerate from a black hole, is it possible light could accelerate to near infinity in very weak or zero gravity? Note in the figure that our laboratory for making such observations which is our solar system is relatively small. Gravity strength changes over a very wide range. We only know it in a very small range.

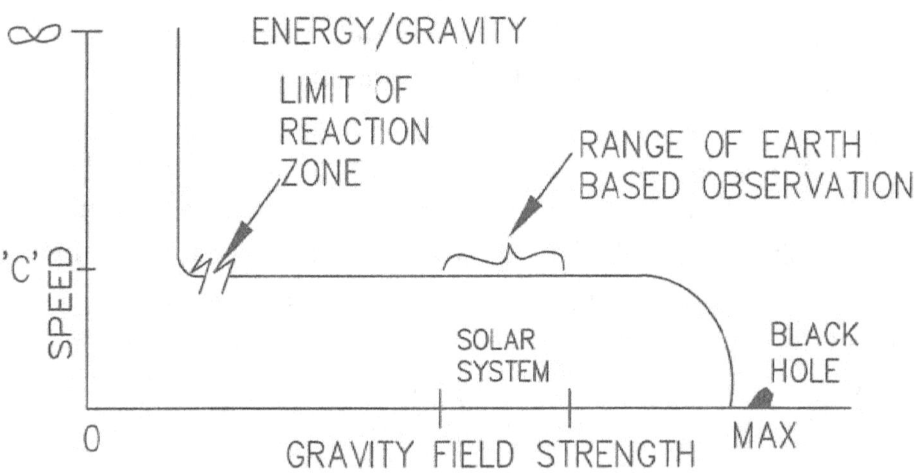

Figure 4: Limits of gravity influence on light

We already know from measurements that light changes speeds in different mediums such as vacuum, air, water and glass. What we do not have, are rules for determining what happens to light in the two extreme conditions as shown above, either in black holes or in zero gravity.

Measuring Distances and Mutual Gravity Attraction in Space

Farthest Known Galaxy in the Universe Discovered

https://www.space.com/18502-farthest-galaxy-discovery-hubble-photos.html

A new celestial wonder has stolen the title of most distant object ever seen in the universe, astronomers report. *The new record holder is the galaxy MACS0647-JD, which is about 13.3 billion light-years away. The universe itself is only 13.7 billion years old, so this galaxy's light has been travelling toward us for almost the whole history of claimed space and time.*

The image of this galaxy is nothing more than a smudge so there is nothing to see that would add to the subject.

Let's consider our best telescopes aimed at any galaxy that is claimed to be about 13 billion light years away. We cannot see it directly, but by today's cosmology theories, we see something that is claimed to be in existence 13.3 billion years ago. Considering the vast distances and so much interference using the latest methods known today, it is impossible to know what we are seeing for sure.

Remember the formula: $(s=d/t)$. Speed is equal to distance over time. We claim we know the speed of light, but we do not know the distance or the time. We can only guess at distances based upon red-shift. Red shift only tells us an object is moving away. We will need to look into the reliability of red-shifts as a measure of distance.

There are many factors that go into red shift measurements of distant objects. It is safe to say that red shift measurements are still in the early phases of development. In the context of this work, we will accept the basic understanding that objects moving away shift light toward the red spectrum.

Lets review how the above article has determined the age of that galaxy: from the article, all calculations were based upon red shift of light waves that showed the particular galaxy must have been traveling away from us at a much faster rate than others because it has greater red shifts. In fact, using red-shift scales, the galaxy is moving away at a speed greater than the speed of light! [29] This is strange but is not the point here yet.

The article relies on spectroscopy of its light properties which agrees with or is similar to light of other galaxies. The article also depends upon light magnification caused by other nearby galaxies claimed to have amplified its light so it can be seen by our best telescopes.

Because of its distance and low light intensity, without magnification, we would not be able to see it, even with our best telescopes. We will get into how this magnification is claimed to occur. We can certainly conclude that measurements across the Universe are very complicated and depend upon many factors. So many different things can happen to light crossing the Universe. In this case, all that can be seen is a blurred image of something behind other galaxies that are obstructing and refracting the very distant light.

This current age and distance technique method relies in part on the Doppler Effects or red-shift which tells us that this galaxy must have been traveling away at a very high rate of speed, for how long? This estimate assumes this particular galaxy must have been in full development 13.3 billion years ago. It says nothing about how or when it was formed. We do not actually see various stages of galaxy formation. Cosmologist claim nearby galaxies are fully developed because it did not take that long for the light to reach us. It is only the claim for how long it took light to reach us that supports the theory that the Universe must be old. If all of the galaxies were created at the same time, they would look exactly like what is seen out there today. All we need to do is show how light crossed the Universe much faster than claimed.

Other than a few anomalies, most everything we see in deepest space is the same as seen in nearby space within our neighborhood of galaxies. This says nothing about the ages of galaxies, for all we know with the information presented, the universe and the referenced galaxy could have an infinite age, even much longer than science speculates or, they may be only six-thousand years old. Red-shift measurements is a new and complicated science for observing such distant objects. So much so, that cosmologist have had to claim that distant galaxies are moving away at speeds much faster than the speed of light.

Is this discovery a result of "odometer" error?

This begs the question that if the Universe formed 13.6 billion years ago from a singularity explosion or rapid expansion, the subject galaxy could not possibly be at that distance by today's measurement standard of red-shift. However to get around this dilemma, scientist claim this galaxy must have formed in only 0.3 billion years. They also claim that the light was somehow magnified by lensing which we will get into later. We cannot really know or calculate galaxy distances but we can reasonably claim it is very far away.

Let's say there was a Big Bang and the material of the Universe spread out from it at an initial speed of say *0.5C* or one-half the speed of light. This is an arbitrary number used here to simplify the point but it is a very high speed number for anything of mass. This would mean that the Big Bang material for that galaxy 13.3 billion light years away would only be half way out there yet. You cannot have a galaxy 13.3 billion LY away if the material to make it has not reached there yet.

The most powerful conventional blasts have an initial expansion speed of about 10,100 m/s. Estimates of supernovae explosions are estimated at about 1,000 times faster. In one year, the distance a supernovae could spread materials into the Universe is about 0.0337 LY. This is only about three percent of the speed of light. Physics tells us that the material of the Universe could not have spread out at the speed of light since according to the theory of relativity, there is not enough energy in the Universe to accelerate any matter to that speed. How can we then reconcile that distant measured galaxies can be 13 billion LY away but the materials needed to make them can only be a very small fraction of the way there? They claim the Universe itself must be expanding carrying the galaxies with it in order for the red-shifts to make sense. That's like saying the road I am driving on is also moving away from me so I am actually traveling much faster than sixty-miles an hour!

The current expansion estimates of the Universe could be nowhere near the speed of light and there is no known way to determine if the Universe expanded faster in the past than it is now or if it is speeding up, or if it is expanding at all. So to get around this physics problem, cosmologist have decided the space the Universe is held in is also expanding. Debates exist over whether the Universe is expanding. The expansion rate is either increasing or decreasing depending upon how changes in red-shift distance measurements fit with their theories. They presently claim the expansion is greater than the speed of light! Of course it could be that something is wrong with red-shifts.

If the Universe is in fact young, then we must show the time for light to travel is not restricted in gravity-free space. That impacts distances based solely upon red-shifts and certainly affects the time for light travel. With that proof, we will need to figure out what is really happening with red-shifts.

This we know. We have only observed the Universe for a few hundred years. So far, we have put forth enough information to conclude that there is no *factual* or provable age for the Universe. So the age of the Universe could be considered to place somewhere on a sliding scale. We will be looking into placing that initial time of Creation at six-thousand years to see what happens. We know the Universe is here because we live in it. Based upon the above questionable logic for the age and expansion of the Universe, we simply cannot rule out other possibilities such as Creation of the cosmos about six-thousand years ago.

Remember a gravity wave is a way to measure the speed of gravity which has been found to be the same as the speed of light. The following may sound technical, but will be explained.

https://en.wikipedia.org/wiki/Speed_of_gravity

*The speed of gravitational waves in the general theory of relativity is equal to the speed of light 'c' in a vacuum,.[2] Within the theory of special relativity, the constant 'c' is **not exclusively about light**; instead it is the highest possible speed for any **interaction** in nature. Formally, 'c' is a conversion factor for changing the unit of time to the unit of space.[3] This makes it the only speed which does not depend either on the motion of an observer or a source of light and/or gravity. Thus, **the speed of "light" is also the speed of gravitational waves and any massless particle.** Such particles include the gluon (carrier of the strong force), the photons that make up light (hence carrier of electromagnetic force), and the hypothetical gravitons which make up the associated field particles of gravity (however, a theory of the graviton requires a theory of quantum gravity).*

*In the relativistic sense,[1]as **the gravitational wave, …. , is the same speed as the speed of light***

Notice in reviewing the above, by definition, "c" which is the speed of gravity waves and light is the highest possible speed for *any interaction in nature* – but that determination must be in the presence of gravity.

Postulate E: Energy, including light is massless and must react with gravity in order to be seen or measured. Stated another way, massless energy such as light would not be seen or measured in the absence of gravity.

By showing the speed of gravity equals the speed of light, we have established a basis or postulate that light (energy) reacts with gravity and may only be visible within fields of gravity of adequate strength.

We could reasonably ask "when does a body in space acting with another body in space become so distant that it can no longer act on the macroscopic scale and they become negligible influences on each other? In open space at great distances, massive objects continue to react with each other, this seems to be the case in our very small lab we call the solar system. Remember that gravity field strength decreases exponentially as the square of the distances increase. As stated above although mathematically gravity may never reach zero, it easily becomes ineffective as distances increase when acting on very small bodies. So how far out has gravity reached? Well that totally depends on the age (t) of the Universe or just how far the "bubble" has expanded.

Universal Gravitation Attractions

Can you imagine the true magnitude of gravity? It is so strong, it can apply a force of one hundred billion tons on an object one billion miles away. So how far out into space do we need to go to escape the gravitational force of our sun upon us? If there were one hundred Jupiter's in the same orbit around the sun, would it weaken the sun's gravitational attraction force? No, the sun would continue to mutually attract each Jupiter and each new Jupiter would attract the sun!

Gravity holding what together?

Let's put some thought into the following claim that gravity is holding this cluster of galaxies together.

Figure 5: Clusters of galaxies in deep space

> *https://www.space.com/galaxy-cluster-recordings-eerie-sounds-hubble.html?utm_source=sdc-newsletter&utm_medium=email&utm_campaign=20190310-sdc*
>
> *The website "Space.com" claims The <u>galaxy cluster </u>in question, known as RXC J0142.9+4438, **contains thousands of galaxies held together by gravity**. In addition, each galaxy is home to countless stars — some of which shine brightly in the foreground of the Hubble image.*

The figure above claims that gravity is "holding" together the galaxies shown in the photo. If red-shifts show these galaxies are all of the same distance value, that is, they are all moving away from us at the same red-shift value, there is no known way to factually determine any influence caused by gravity between them. They are moving away because red-shifts say they are moving away and that has nothing to do with any potential mutual gravity attraction to each other.

This figure claims these galaxies are held together by gravity. We cannot look at this photo except in a planar dimension but it is deep three dimensional space. Space is extremely deep, so taking that into consideration, pick any one of the spiral galaxies in the photo. If it is similar to our own Milky Way, it is approximately one-hundred thousand light years across (yes, those tiny disks are enormous).

Our nearest neighbor, the Andromeda Galaxy is estimated to be '2.537' million light-years away from our sun or our Milky Way galaxy. That is about twenty-five Milky Way galaxy diameters away. There are claimed motions between us and Andromeda as measured by red/blue shifts of light. One such calculation not shown is Andromeda will move toward us about 0.484 light years within a time frame of one million years. That's less than one-half of only one of the 2.5 million light-years away or less than two ten-thousandths percent of its distance from us. It is a calculable relative motion based upon Newton's law of universal gravitation.

Figure 6 below states that Andromeda will collide with the Milky Way in about 4.5 billion years. So, yes, mathematically speaking very old massive objects in space could have mutual attractions toward each other over very great distances but do they really? We cannot prove one way or the other, as we only have red-shift to suggest relative motions to each other. There can be no mutual attractions of course, if these galaxies are outside each other's the six-thousand year bubble. If that is the case, there is no mutual attraction. There would be no way to know why Andromeda is showing blue shift relative motions toward us, especially if the Universe is only about six-thousand years old.

Andromeda–Milky Way collision - Wikipedia

*Andromeda's tangential or sideways velocity with respect to the Milky Way was found to be much smaller than the speed of approach and therefore it is expected that **it will directly collide with the Milky Way in around four and a half billion years.***

Figure 6: The Law of Universal Gravitation

Note that "mathematically" in much less time of the claimed age of the universe (13.6 billion years) or a little over 4.5 billion years, the two galaxies would collide. Also, if the universe is expanding, why then is Andromeda moving toward the Milky way? We are not seeking those answers here and not sure we would find them. What we have discussed here *assumes* the Universe is old but our premise is that the Universe in only about six-thousand years old. All we really have to go on is red-shifts.

Figure 7: Andromeda, our closest galaxy

Red-shift motions claim Andromeda, in about 4.5 billion years, will collide with us. There is no known explanation as to why Andromeda would be moving toward us if the Universe is expanding. We can imagine there could be dark matter causing currents or eddy's in the Universe but to date that is impossible to know, or it could be an act of God. The goal of this section is to show that the Universe is enormous and mutual gravity forces may not exist dependent upon the age and size of the Universe.

If the Milky Way and Andromeda could collide in about 4.5 billion years, why would they not have already done so? At least shouldn't we see the spirals stretching toward each other caused by the effects of mutual attractions. Since the universe contains billions of galaxies much closer to each other, should we see signs of distortion toward each other. If that is the case with gravity, then the Universe should look much different with not just a few, but millions of galaxies distorted and colliding into each other.

This certainly builds a case for a young created Universe since gravity fields would not have reached nearby galaxies causing them to attract each other. Logically speaking, cosmology leaves many open questions about our beginnings. Of course cosmologist will claim that due to the expansion of space itself, galaxies will not collide. There is no proof of expansion but for very complicated red-shifts.

Conclusion

Determining a young age of the Universe changes everything and at the same time changes nothing. So far, we cannot rule out either a young Universe or an old one. Everything in cosmology science assumes the Universe is billions of years old. No consideration is given at all for a young Universe.

Chapter 6

Properties of Light

What is a photon?

We are now going to get into the actual physics of energy and light as it is affected by gravity. The photon has a very short life span. As we will see, the photon seems to be the result of the actual conversion of massless energy into something that is capable of doing work. This definition implies that energy in the form of light that is not in the presence of gravity can do nothing. When you feel the heat of the sun on your face, that is energy being converted by gravity into useful work. So photons appear to be energy being put to work.

Lets simplify what is to follow. The postulate that follows states that gravity must be present for energy to do anything. We earlier discussed that gravity is still reaching out throughout the Universe so there remains vast areas in space where gravity does not exist. So energy and light traveling through empty space cannot react with gravity since gravity is not there yet. We need to find out what happens to energy and light while traveling in empty space outside of the bubbles.

Postulate F: Units of light called "photons" may be defined as the product of energy in the presence of gravity. It is an unexplained occurrence or phenomena when combining massless energy with gravity. This combination produces harmonics or vibrations we understand as light waves. This phenomena converts non-material massless energy into useful work (or photons) of which life is dependent upon. A photon may only exist within a gravity field of sufficient strength.

As a follow-on, this energy conversion phenomena is not limited to the visible spectrum of light as all forms of energy must be capable of being identified, measured and converted into useful work. We must also consider that in order to react with energy, there must be sufficient gravity strength capable of converting that energy into work. This leaves open the question of how can light leave one gravity field and enter into another.

Massless particle – Wikipedia
https://en.wikipedia.org/wiki/Massless_particle In particle physics, a massless particle is an elementary particle whose invariant mass is zero. The two known massless particles are both gauge bosons: the photon (carrier of electromagnetism) and the gluon (carrier of the strong force). ... Neutrinos were originally thought to be massless.

We need to question exactly what a photon is and if it can exist outside of gravity. The answer appears to lead us to think of energy in zero gravity as simply energy in its purest form and perhaps in some unrecognizable or undetectable state that does not contain photons or possibly even wave forms. The following article provides support as it postulates that light can behave in different ways depending upon its environment.

---

https://www.scientificamerican.com/article/space-based-test-proves-lights-quantum-weirdness/

Scientific American Article:

*Physicists sometimes say that a beam of light traveling through space is like a "great smoky dragon." **One can know much about where the light comes from (the dragon's tail) and where it is seen (the dragon's head), yet still know precious little about the journey in between** (the dragon's mysterious, nebulous body). **As light travels from source to detection, it can behave as either a particle or a wave—or, paradoxically, both states or neither state.** Now an experiment using laser beams shot at satellites in low-Earth orbit has confirmed that this bizarre detail about the nature of light holds true across record-breaking distances.*

*Quantum physics, the best description yet of how all known particles behaves, suggests that reality is fuzzy and uncertain at its most basic levels. For instance, **the surreal quantum effect known as superposition essentially allows electrons, atoms and other building blocks of the universe to each exist in two or more places simultaneously**.*

Another strange quantum phenomenon is particle-wave duality.** Whereas Isaac Newton thought light was made up of particles, his contemporary the Dutch scientist Christiaan Huygens argued that it consisted of waves. Eventually, researchers performing the so-called double-slit experiment demonstrated that Newton and Huygens were both right—**photons of light could behave as both particles and waves.

This article states [light] *can behave as either a particle or a wave—or, paradoxically, both states or neither state.* (particle wave duality)

The article states [light can] *exist in two or more places simultaneously.*

What we can take away from this article is to question light and energy interaction in micro-gravity or zero-gravity fields. We may find that while energy is observed and predictable in gravity, it may not be observed in zero-gravity and may exist in two places at the same time. We are going to apply that logic as we progress. The concept of "particle-wave" duality opens many doors for further thought.

Is light measured to one universal speed?

There are differences in outcome of measuring the speed of light such as differences in vacuum, air, water, etc. The most common way to show that light speed can change is to use two parallel flashes of light, one in air and one in water: the light flash in air would be seen at the test point before we see the one through water. Why is it seen at different speeds? Because light is dependent upon gravity and mass density of the medium it travels through. This is important later.

Considering particle physics, can light consist of any matter including sub-atomic particles and still reach the speed of gravity? Not according to Einstein's physics. Where can we measure or perform tests on the speed of light? We can only measure the speed of light in gravitational fields.

I have always been perplexed by a gut feeling about relativity as it has been explained describing time as either speeding up or slowing down depending on the motion of the observer. I have read about results proving relativity using particle physics in controlled test environments. So how could light possibly travel across the universe any faster or slower than the speed limits Einstein's relativity has determined? Let's see if we can answer the question by first unraveling the dilemma.

How is Einstein's Theory of Relativity limited to our environment?

Einstein's theory of relativity only works in gravitational fields as time does not exist outside of gravity. Imagine where you are at this very moment reading this material and imagine I am on a planet one million light years away from you at this very moment (some readers may think that is where I'm coming from). How can I communicate with you? The speed of thought is instantaneous. However, we live in a physical world dependent upon physical properties to communicate. Even though we can think the same thoughts simultaneously, our thoughts do not depend upon physical properties as we know them.

If there existed a non-physical medium outside of time, call it a fabric, an ether, dark matter or a "non-gravitational medium", then maybe massless energy could travel through this medium virtually unconstrained until it reaches the physical realm of a gravitational field where it slows down to the speed of light. Since science knows nothing about any other possible "mediums", gravity-free space becomes our non-physical medium. The main difference between our realm of instantaneous thought travel with respect to light travel is the interjection of time and light being constrained by gravity. More will be said on this later.

We may also consider that if light requires gravity to organize itself into waves, both must work together but with exceptions. The intense force of gravity is claimed to not let light escape a black hole. If a black hole is still active and producing energy and light, then what is the speed of that light if it cannot escape the black hole? The speed must be zero if light cannot escape. This implies that gravity may compress so densely around a black hole that light generated within the black hole is blocked or stopped by its density. At the other extreme, as gravity becomes so weak, light may escape gravity and some new set of rules may apply. We saw this in *figure 4* earlier.

Lets reference the following very interesting but technical paper for some other possibilities with the speed of light. This is very important in our study. The following is not the actual paper presented by Mr. Shirazi. It is a partial summary referencing his paper. His paper opens doors, so please review it for the comments and further use in our applications.

A Derivation of the Speed of Light Postulate [24]

by: Armin Nikkah Shirazi

University of Michigan
Ann Arbor, MI 48109

November 28th, 2008

Note that Slayton's comments are all in type while Mr. Shirazi's comments in this paper are in *italics*. **The paper itself is not provided here** but is readily available on the internet at the referenced link. The paper introduces some rules that we observe in our studies of physics. Mr. Shirazi states *"The speed of light is an assumption upon which Special Relativity is based."*

For the layman, do not let the following explanations confuse or discourage you from this work, as it is mentioned only to present paradoxes in logic and reasoning in contemporary theories. Mr. Shirazi's work is profound and exceptional in leading to new ideas.

Mr. Shirazi presents three axioms (postulates) as a starting basis.

1. Motion in proper time (Newtonian mechanics)

2. The nature of "existence" of entities which do not age (using Special Relativity)

3. An axiom which defines "existence" in a space-time to be transitive. (Meaning it can come and go in and out of existence per Special Relativity)

The following are excerpts from Mr. Shirazi's paper.

"The second postulate ('speed of light postulate'), given that it applies to a finite speed, **still seems just as mysterious today as when it was proposed over a hundred years ago because it seems to defy our intuitive ideas of motion.** *Any approach that could be used to resolve this apparent mystery will be based on a certain preconceived notions, or biases, and it is best if these are clearly identified in the beginning. The principal biases that underlie this paper are the ideas that the laws of nature do not violate the laws of logic, and that it is within our ability to*

comprehend them. From the first bias follows the view that there **must be a logical explanation** *which can explain why this postulate is correct. From the second* **it follows that such a logical explanation should also be understandable, i.e. physically make sense to us."*

The application here relating to this work is to provide a logical explanation that is understandable and makes sense to us. He goes on;

"The existence paradox, then, is this: How is it possible that entities traveling at the speed of light 'observe' their own duration of existence to be precisely equal to zero, and yet produce observational consequences evidencing that they do exist? Notice that if we lived in a Universe in which universally $c = \infty$, then this apparent paradox would not occur:"

This paragraph seems to imply that anything traveling at the speed of light using special relativity would observe the time of its existence to be zero. They do not exist in time. Then it goes on to say that if light traveled at infinite speed, this paradox would be solved.

It seems, while we are looking at the speed of light, we should actually be looking at something else and we will, the speed of gravity. The paper goes on to explain that since photons of light exist for a duration of zero time for an observer traveling with them, then photons may not exist outside of space/time, if they did, they could travel at infinite velocity in the absence of gravity.

If we substitute photons generated in gravity fields with pure energy traveling outside of space/time which is outside the effects of gravity, we can begin to see a plausible and logical explanation for how light could change properties and traverse the Universe much faster than the classical view of special relativity.

Remembering our premise that gravity is still in the state of expansion as the "bubble" concept so stated earlier. Energy then, must also exist "outside of gravity-effective" fields but in some transitive state that does not obey laws of physics that require gravity. If there is no gravity of effective strength, then how can energy be detected? We must consider whether zero or micro-gravity fields somehow lack properties necessary to react with energy to produce light. Our goal is to get there in this work.

In Mr. Sharazi's paper, he made the following statement regarding the existence paradox: *"A problem like this sometimes requires a profoundly new idea, highly unfamiliar at first, after the introduction of which the resolution of the problem becomes almost trivial."* The purpose of this work is to introduce new ways for others who are more capable than me to solve the riddle of light travel across a young Universe. All effort employed here is to look at our beginnings differently. Lets work from the premise that the Universe came into existence about six-thousand years ago, and see what happens. This work attempts to establish the ground work or building blocks for future study.

Energy and Light Velocity Source-to-Observer Postulate

The following postulate is not this writer's, it is a universally recognized postulate for the speed of light.

> ***2nd Postulate of Relativity*** [32]***: The speed of light in a vacuum is the same for all observers, regardless of their motion relative to the source. Implications: The speed of light is a Universal Constant.*** *Feb 18, 2006*
> *www.astronomy.ohio-state.edu › ~pogge › Ast162 › Unit5 Lecture 31: Special Relativity Search for: How is speed of light same for all observers?*

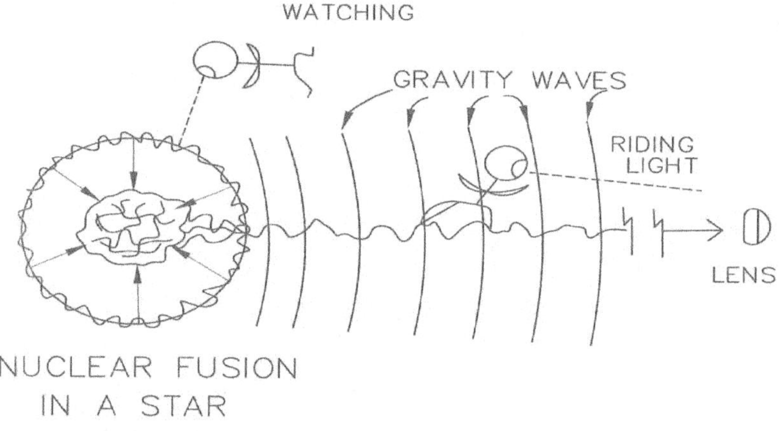

Figure 8: Gravity organizes light into wave form

How is it possible the speed of light can be the same for all observers? Referring to *Figure 8* above, lets see what happens to light and observers. We are going to follow an entity traveling with light from the moment of production inside the nuclear reaction of a star. The process begins as fusion-able matter like hydrogen being converted into massless energy. We place our entity beside the reaction to witness what happens. There are billions of these reactions happening each in their own time but not in sequence or harmony, so the products of energy produced is just an unorganized form of noise. By some phenomena, the energy must first organize itself into wave form. Gravity is the only recognized agent that can organize this energy noise into vibrations as wave forms. After formation into wave form, the entity attaches himself to the energy and travels with the light and the expanding gravity field as it travels outward at the speed of light. Any attempt to measure its velocity conforms with the recognized and accepted speed of light which is the same as the speed of gravity waves. So far, we are in full compliance with Newtons laws and special relativity for the speed of light.

Gravity must necessarily have been the force that propelled the newly produced energy and rider from the moment of its creation to the speed of light and organizing it into wave forms. All available light and energy wave frequencies are then viewable in their respective spectrums.

Radio Wave Analogy

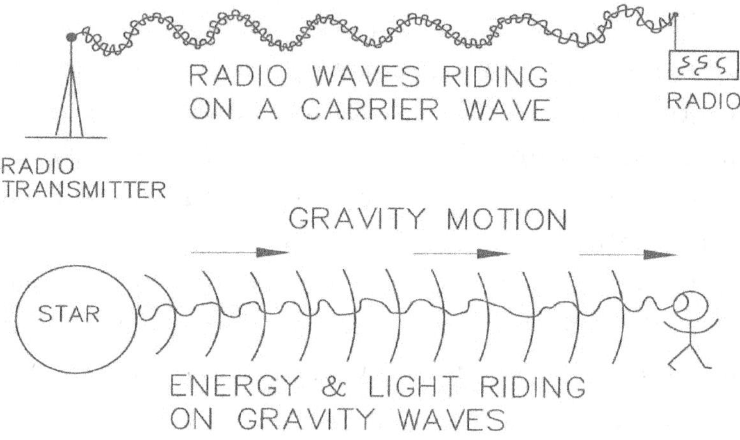

Figure 9 Light and Energy riding on Gravity waves

Referring to *Figure 9*, we use an analogy to see what is happening to light as it travels from the star. We learned something about electro-magnetic radiation from our history with radio waves. Radio waves [26] can be carried within a specific man-made radio frequency outward at the speed of light. That carrier frequency then contains within it, broadcast signals that are converted into the voices and music we hear on the radio.

For purposes of understanding how energy and light are transformed from noise in a nuclear reaction into waves traveling at the speed of light, we can think of light as radio signals but riding on gravity waves. Gravity must be the natural force, or carrier of all electro-magnetic energy radiation traveling outward from nuclear reactions. Since gravity is capable of organizing and carrying all forms of energy, it plays a much greater role in physics than just a force holding us down to the ground. Again, we use radio signals here as an analogy for what happens when energy attaches itself to gravity waves and ride outward at the speed of light.

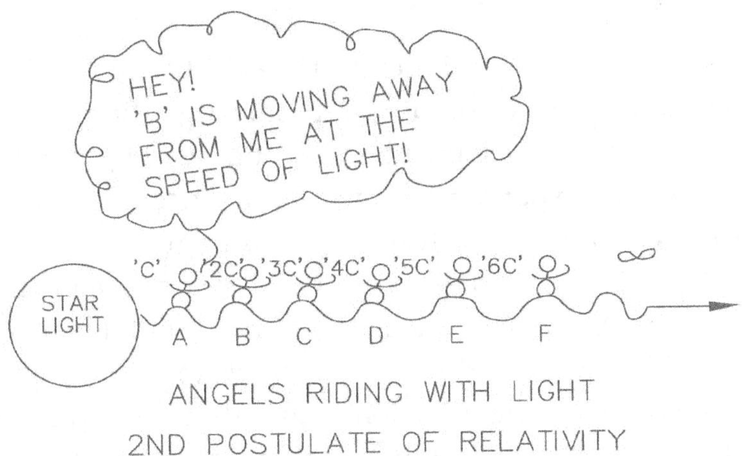

Figure 10: Angels playing around on light waves

Referring to *figure 10*, imagine a massless observer jumping aboard a unit of light. Then another observer jumps aboard another unit a minute later then another and so on that each observer is riding on a unit of light about one minute apart.

To make this interesting, lets say these observers are *massless angels* riding on the light. Angel 'A' is traveling at the speed of light and looks to see Angel 'B' ahead of him. The 2nd Postulate of Relativity states: *The speed of light in a vacuum is the same for all observers, regardless of their motion relative to the source.* The speed of light is the same for all angels. Angel 'B' is also traveling away from Angel 'A' at the speed of light from his point of observation but he is already traveling at the speed of light. So Angel 'B' must be traveling twice (2C) the speed of light. Then, Angel 'B' looks ahead at Angel 'C' and sees him traveling at the speed of light away from his unit of light, so Angel 'C' must be traveling three times (3C) the speed of light.

This process can continue until the light speed of the angel in the very front could be approaching infinity. Obviously at this speed, the massless energy of light will escape the star's expanding gravity field. Once the speed approaches infinity and is traveling in zero gravity, it will remain traveling at infinite speed until it reaches another gravity field. Once there, the light will react with the new gravity field, slow back down and convert to visible light. It will travel opposite the direction of gravity wave expansion but still at the speed of gravity to a ground observer.

So light 'energy' can actually exist in two places at the same time as stated in the Scientific American article and as I believe Mr. Shirazi's paper shows above. Light travels both at infinite speed independent of gravity and also responds to gravity and appears to ride with gravity regardless of the direction of gravity expansion. The light riding with gravity is converted into photons producing visible light.

Within the physical Universe where gravity is present, a ground observer will only see the light responding to gravity at the speed of light. Even though light and energy are accelerating to infinite speeds, all we can see is light reacting with gravity being converted to photons.

Postulate G: Energy and light from the moment of production within a nuclear reaction accelerates radially outward at infinite velocity. However, energy and light can only be observed and measured at the speed of gravity within a gravity field.

The paradox is that we can only see energy with respect to what gravity allows us to see at speeds established by the Creator. Because we live in the reality of a physical universe, we are restricted to what we can see. We cannot see pure massless energy. Pure energy can transcend space and time by traveling inside and outside of gravity at infinite speeds. Looking at it in another way, gravity is the carrier frequency like a tuned radio which is only able to tune in to whatever it can on its natural frequency. So gravity is the natural carrier of visible light and all other detectable energy forms.

To help clear our understanding, lets look again and follow step by step, light produced in a star at 13 billion LY away and coming toward us. We will again use the angels who emerge from the star's nuclear reaction as they travel toward to our location near our sun.

Step 1: As seen in Figure 8, the angels observe pure energy being randomly produced as noise inside the star's nuclear engine. The gravity present inside the star organizes the energy into vibrations and accelerates it outward.

Step 2: The angels one by one, grab on obeying the 2nd *Postulate of Relativity* and accelerate step by step up to infinite speed. Note that all of the available types of energy in that star are organized into their respective wave-forms by the carrier which happens to be gravity.

Step 3: The acceleration of the angels and energy created in the star almost instantly passes ahead of the star's gravity influence. The angels are traveling at infinite speeds and the energy is undetectable and incapable of doing any work in gravity-free space.

Step 4: The angels and energy from the star arrives at our solar system and our sun's expanding 6 KLY gravity field. The Sun's gravity field detects the energy just as a radio would detect a signal that it is tuned for, so the energy once again becomes detectable in our solar system. The visible energy then travels toward us at the speed of light through our 6 KLY expanding gravity bubble and we see it six-thousand years later.

Step 5: Now we must measure the time and distance the angel traveled to reach us. First, the distance is not in dispute however, the accuracy of the distance is based upon a few assumptions since the distances cannot be precisely measured. So let's stipulate that the distances are very near what astronomy and physics claim.

Determining the time: The massless energy and light accelerated through its 6 KLY expanding gravity field and quickly approached infinite speed from the moment of production. The light immediately crossed gravity free space until it reached our sun's gravity field which is also 6 KLY. When the light reached our gravity field, it responded to gravity and produced visible energy and light.

To solve the time is simple, we apply (t=d/s) so we have 6 KLY divided by C which is the speed of light and that equals six-thousand years. Even though the star is about 13 billion light years away, its light or the angels would take six-thousand years to reach us if they had left that star today.

It follows that if light left the star six-thousand years ago at the time of creation, it would have reached us immediately as our gravity field was just beginning to expand. We can only measure the speed of light within gravity fields. If the light left the star one-thousand years after Creation, it would have taken one-thousand years to reach us. Another important observation is that from the moment of Creation, all stars in existence would be seen immediately but as time advances, as in our case above, any light leaving a star today would take six-thousand years to be seen.

The Dual Properties of Energy and Light

Figure 11 shows how light responds to the 2nd *Postulate of Relativity*. The dual properties of light allow it to be seen and observed within its local gravity field and also allows light to accelerate up to infinite velocities.

In this case, the light from the distant star acted in both modes by traveling toward earth at infinite speeds, then slowing to 'c' when reaching our sun's gravity field. Light travels locally at the speed of gravity when in gravity. Light also accelerates to infinite speeds in its dual role.

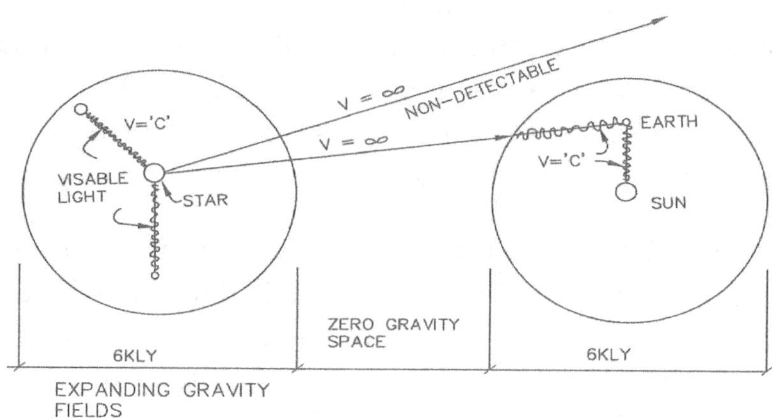

Figure 11: Speed of light in gravity-free space

Postulate H: The initial velocity of massless energy radiation is equal to infinity. The velocity of "usable" energy instantly changes to the speed of gravity in the presence of any new gravity field.

It is interesting to note that light arriving at our expanding gravity field is traveling opposite the direction of our expanding gravity field yet its observable speed is the same regardless. This supports the 2nd *Postulate of Relativity* for the speed of light regardless of relative motion.

Conclusion

This chapter has opened the idea that strange things happen to energy and light as it can exist in two places at the same time and it can travel at infinite speeds when in its pure energy state. Gravity is the carrier of visible light. We have seen two significant ideas, about what happens to energy outside of gravity and how gravity must be present for energy to do any work. We speak of energy knowing that light is part and parcel of the energy spectrum. We are not done with this study, so lets move forward with more explanations.

Chapter 7

Light Physics, Light Lensing and Time Changes

There are three properties of light behavior we will look at beginning in this chapter. In the most simple terms possible, this chapter will explain what happens to light in space and how it can affect time. We will later see how this chapter has an impact on the way distances are measured in space. We will also look into the mystery of red-shifts showing galaxies traveling at greater than the speed of light. Lets quickly review the terms we will be using.

Refraction: The bending of light when it passes from one *medium* to another at an oblique angle. The changes in *mediums* can include air, water, glass, gravity and zero gravity space.

Dispersion: When a beam containing more than one frequency such as white light is split into a number of different beams. A prism is commonly used to separate light beams.

Diffraction: Light waves bending around the edge of an obstacle. An example would be a light bulb in a bedroom casting dim light out into a hall where the bulb is not seen.

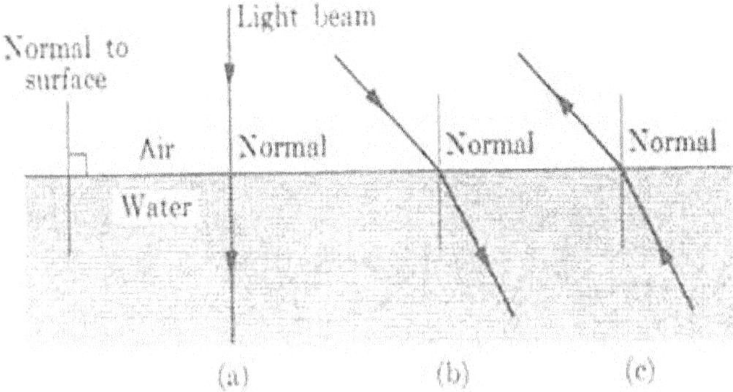

Figure 12: Refraction of Light

Referring to *Figure 12*, the bending of light when passing from one medium to another is called *refraction*. Physics experiments show that a beam of light passing obliquely from one medium to another is deflected at the surface of the medium. When light is deflected, it is toward the "normal" or *toward* the perpendicular to the surface of the medium. If light then passes from the medium back to the original, it deflects *away* from the normal. We can witness this effect in space as light from one star passes through the gravity field of another star. The bending of light passing near a star in space is claimed by cosmologist to bend due to following curved gravity around the object. Or, light obeys the nature of light refraction by bending when passing through different mediums. Both explanations are acceptable because light does bend. The change in medium in this case is when light coming from one star enters a gravity field of another star.

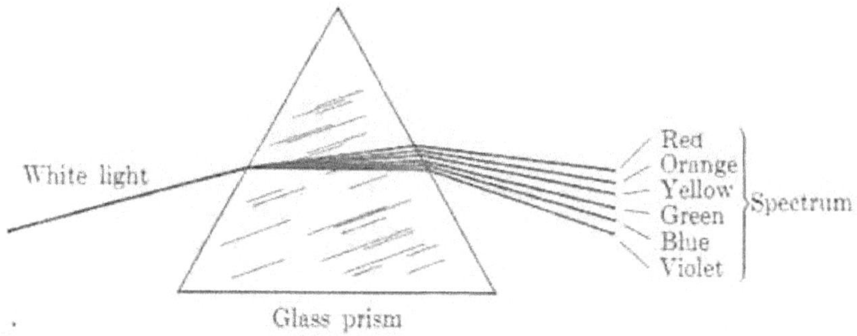

Figure 13: A prism separates light of various frequencies

Dispersion of light will prove very interesting especially when light eclipses the gravity field of another star. Due to the curvature of the gravity bubble, light passing through can produce an arc of light we call lensing.

Diffraction of light is produced by electro-magnetic radiation of energy converted to useful work in the presence of gravity. That's why a lightbulb in a room can cast light outside a room that is not in direct view of the light. This principle of diffraction causes light to lose energy through radiation within gravity fields and other mediums such as air. When massless energy is traveling in zero gravity, it is not being converted into visible light so it does not lose energy through diffraction.

Refraction of light as it relates to changes in medium densities

Consider a clear glass ball or a glass of water with laser light applied to it from one side and the light passing through it is displayed on a white background on the other side. If the light passes through the center, and is perpendicular to its surface, there is little or no refraction (or bending) due to there being no oblique angle to cause refraction. As a beam is moved outward, the light begins to refract. The light first refracts *toward* the normal and then as it exits, it refracts *away* from the normal. Figure 14 shows how this light changes position on a background surface.

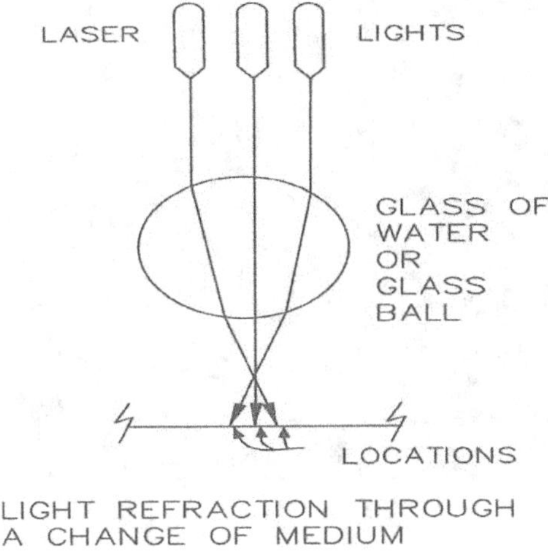

Figure 14: Refraction bending light through a change of medium

Note that on the wall of the figure above, three separate beams of light apear to cross each other. It is interesting what an observer would see if positioned at the wall.

In this simple experiment, as a beam of light would move from the outer radius toward the center of the glass, the beam will center appearing as though it did not pass through the glass at all. An oblique angle such as the curving surface is required to cause refraction.

One star in multiple locations

Is it possible that sometimes two or more stars we see are actually the same star? Referring to Figure 15, light from one background star may refract and disperse its light around another star in a way that one observer in the proper spot could see the light appearing as multiple stars. *Einstein Cross* is an example of this light refraction phenomena.

Figure 15: Einstein Cross

Just as light passes through the water glass in the earlier example, the same thing happens when light passes through a gravity field of a star or a distant galaxy as in Einstein Cross. The difference this time is that rather than a beam of light being moved, it is the position of the observer. We cannot move the stars but we can view the results from various positions.

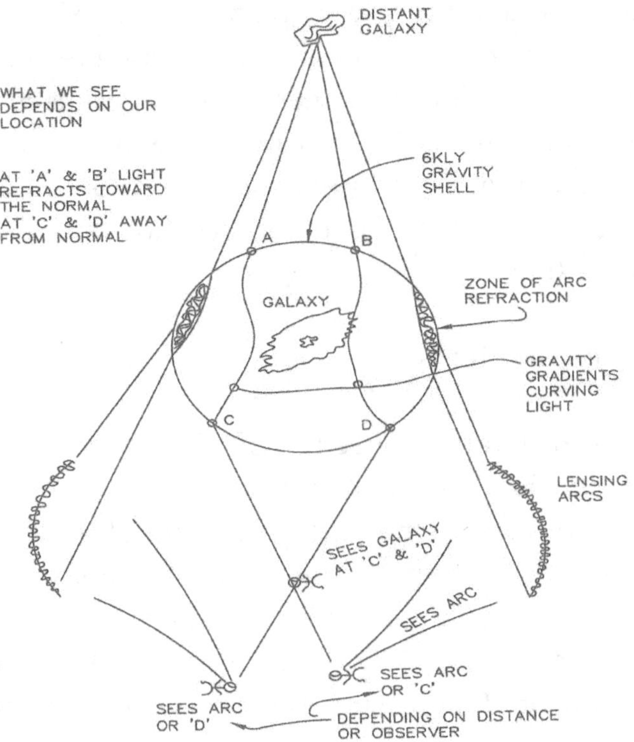

Figure 16: Light Passing through a Galaxy

As seen in *Figure 16*, an observer located at the intersection of C' and D' would see two images of the same star or galaxy. The two images would appear to be located at positions C' and D'. This drawing is two dimensional, if it were a sphere, then four images would be seen just as we see "Einstein Cross".

In *Figure 16* above we also see what happens to light at the edge boundary of a gravity field. An observer at locations shown would see the light dispersed in an arc or at locations C or D. The distance of the observer from the foreground galaxy will determined what the observers will see. The arcs are seen at the outer band of the gravity field where the background light eclipses the foreground galaxy causing the light to disperse in an arc. The arc is seen by a proper distanced observer formed around the radius of the foreground gravity shell. Consider that if the gravity shells are continuously expanding at the speed of light, observations from earth will change over time.

Gravity Gradients

What is different between a glass ball or a glass of water and a gravity shell around a star or galaxy? The density of the glass ball and water is the same throughout but with gravity, the intensity increases proportionally as we approach the object. So light traveling through a gravity shell experiences continuous changes in gravity field strength. We call this a gravity gradient of increasing magnitude or field strength as we approach the star or galaxy. We can conclude that this gradient is the same as a *continuing change of medium* as light is passing through.

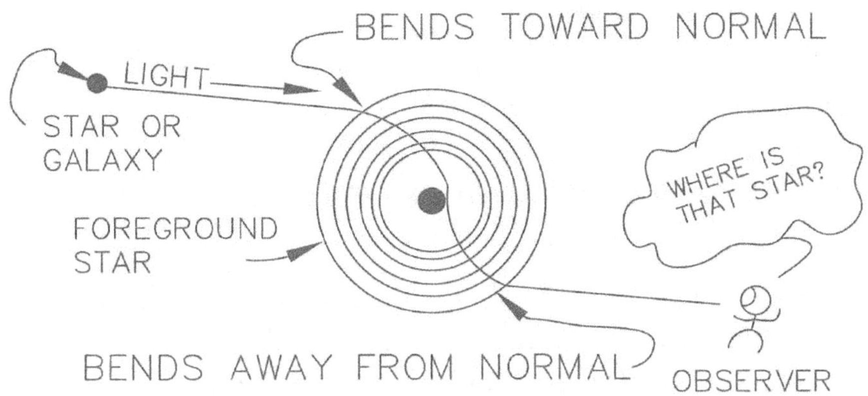

Figure 17: Showing light refraction caused by a gravity gradient

Figure 17 shows that if light refracts at every change in medium, then a continuous change of medium will cause a continuous change in refraction so that a curve is produced. A gravity gradient bends light so that it continually curves inward *toward* the normal then as it exits, it curves *away* from the normal until it exits the gravity gradient.

How does time change between two observers?

Most importantly and always lacking from our studies about time and light in the Universe is knowing exactly when the light under study was initiated. What we typically see from space are streams of light coming to us from distances and times unknown.

Lets see if we can simplify some of the mystery of Einstein's Special Relativity. Don't worry as this is only an overview. Time dilation which is part and parcel to Einstein's Special Relativity is introduced here to point out only one thing. Originally, relativity was all about the speed of light changing time between observers. Since the introduction of the speed of gravity and gravity fields, time dilation is also "*being differently situated relative to a gravitational field.*" We could say that objects in different positions within a gravity field or gradient would sense time dilation. The claim is that as gravity fields change between observers, time also changes. The claim above implies changes in time take place because of something to do with changes in gravity due to gradients of gravity and that is due to *refraction*. Lets see if we can simplify this claim.

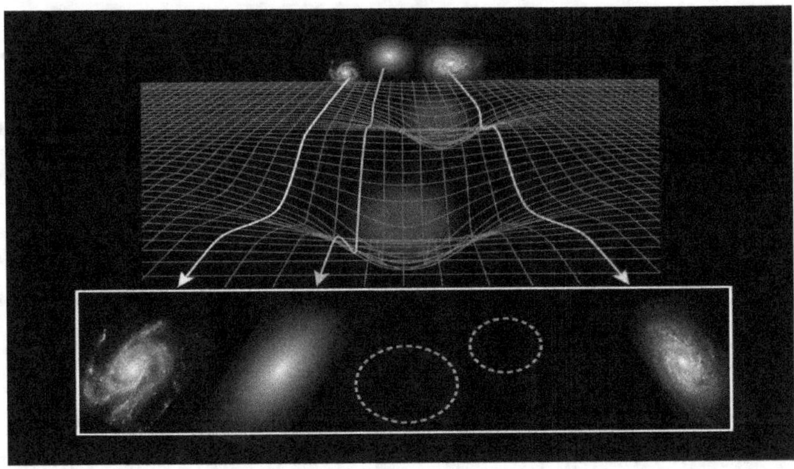

Figure 18: Light refraction caused by gravity gradients.
Courtesy of Physics.org

Referring to Figure 18, It appears that light bent by gravity causes light to travel further distances than it would in a straight line. One observer in a gravity field may see light traveling longer or shorter distances resulting in more or less time travel. Remember that the closest distance between two points is a straight line.

It is claimed that gravity "curves" space but it is actually gravity gradients curving light. If this is the case, then time dilation is simply caused by light traveling a farther distance than it would a straight line and has nothing to do with relativity. It is basic physics.

__Postulate I:__ Since the speed of light in gravity shells has the same value for all observers regardless of their state of motion, the shortest possible time for light to travel is in a straight line. If light follows a longer path such as curving due to gravity refraction, we are simply adding the time necessary to travel a longer distance. Time dilation is nothing more than light traveling longer distances taking longer times to arrive.

We cannot actually see light curving or bending because we can only see the light that has arrived at our location. The actual curving of light due to a gravity gradient causes a time/distance change taking light on a longer path. If the speed of light (or gravity) is constant then using *speed = distance/time* (s=d/t) shows that both time and distance had to change proportionally since the speed does not change. This could cause havoc to the theory of relativity. If the travel distance is longer, the time to travel is proportionally longer as well.

So, does time alone really change or did the distance light traveled on a gravity gradient curve change proportionally taking more time? Figure 18 above shows that if gravity bends light along its path, then it will take a longer time to reach the observer. We can conclude that massless energy (light) must react with gravity following gravity gradients as shown in the figure above. More accurately, light responds to gravity by producing photons with some mass, then photons respond to gravity by curving. Claiming that space is curving light adds to confusion.

New Ways to Measure Time Travel and the Speed of Light

Postulate J: The time required for light of an event such as a supernova to be seen can be derived by (s = d/t) or transposed to (t = d/s). The distance is the sum of distances where light travels through gravity fields: (d_g=d_1+d_2+d_n...). Since 's' is the speed of light 'c', the equations becomes (t = d_g/c).

Light refraction through gravity gradients is one of the primary causes of red-shifts. The number of gravity fields that distant light travels through may be determined by the amount of red-shift. More than likely, star light (or light from a galaxy) that is 13 billion light-years (13 BLY) away is so distant that its light would most likely have to pass through a large number of expanding gravity fields of galaxies and stars. This could curve the light many times adding more time and distance for the light to reach us and that adds to red-shifts.

If we followed the light leaving a galaxy in *figure 19* below and it traveled through several other gravity fields along the way, the observable red-shift and time to reach us is added to by each gravity field it traveled through. However, due to infinite velocity of energy and light when not in the presence of gravity all galaxies and stars would have been immediately visible at the time of Creation. At Creation, there were no expanded gravity fields.

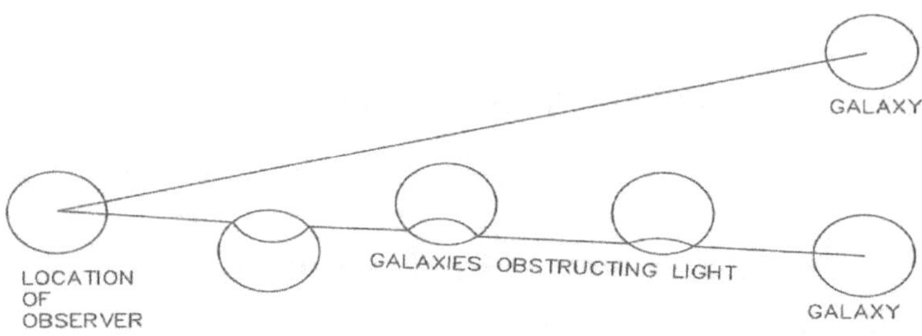

Figure 19a: Light travel through gravity bubbles adds more time and distance

Are any astronomical constants really constant in cosmology? There is contemporary research showing fundamental physical constants, such as G', the gravitational constant, can change slightly. G' is a value assigned for calculating universal gravity attractions of bodies of mass. Evidence for changes in the speed of light is still being researched. This work has shown some possibilities for changes.

An analogy for setting limits on the speed of light is: if you drive from Los Angeles to New York City, it must take you about 50 hours because the speed limit is 65 miles per hour. There is no other way and if you even think you can get there faster, you are mistaken and in violation of all known rules for traveling to New York. With this limited type of thinking, we have eliminated all sorts of other possibilities for traveling to New York such as rail and air travel. Two-hundred years ago we did not know much about rail, automobiles and especially air travel; the same applies to understanding other ways light may have traveled across our Universe.

Remembering that gravity shells or bubbles currently have a radius of six-thousand light-years (6 KLY), we need to look at another possibility for what can happen to light coming from zero gravity and entering a gravity shell. Refering back to *Figure 19*, the light experienced changes of medium from gravity to zero-gravity. We have shown it is possible that its speed could change in and out of gravity just as it does from air to water. If this is the case, it further complicates distance and time measures from space. Considering all of the evidence for what happens to light, it makes sense that if light speed slows down when it enters a dense gravity shell, it may well have been traveling much faster outside of gravity.

Light intensity and Distant Viewing

Postulate K: Light and energy from distant galaxies is refracted and dispersed in accordance with Newtonian physics. The laws of physics in a gravity-filled universe would dissipate energy so much, that distant objects could not be seen. In a young Universe mainly void of gravity interference, light will not refract or disperse as it would in a universe filled with gravity gradients. Light must necessarily travel through gravity-free space conserving energy in order to be seen.

This postulate is an empirical review of facts we have studied previously. As we look into deep space using our best telescopes, we see so much homogeneity since distant galaxies appear so similar to our own. This similarity transcends observed distances out past ten billion light years and more. We do not actually see aging of the Universe as we should expect because we are claiming we are looking back in time those billions of years. What we actually see is they all appear to look the same or about the same age.

Looking at the study of light intensity in physics, the ratio between the speed of light 'c' and its speed 'v' in a particular medium is called the index of refraction of the medium. The greater the ratio, the greater the refraction causing light and energy dispersion. If light energy is dispersed over billions of years in space affected by gravity, the dispersion would completely dissipate its energy and be unseen.

Log scales [36] used for measuring light intensity from space

Logarithmic scales are used because the actual light intensities cover such a large range, it would be difficult to represent them on a linear scale. Our eyes respond to light in a logarithmic way and see each doubling of light intensity as equal changes.

We have seen how light disperses as it travels throughout the Universe. We also know that it takes very deep time exposures to get any viewable results from distant galaxies. We know that light diffuses, disperses and refracts causing it to lose energy as it travels through gravity fields. What do we know about what happens to light as it traverses space outside of the gravity bubbles? If it can be shown that light travel in gravity-free space is much more efficient as there is no interaction with gravity causing it to lose energy, this may explain how we are able to see light from such distances.

We earlier studied that light actually bends as it travels through gravity gradients. This change of medium is causing dispersion and refraction. In an old Universe, light travel reaching us would have been passing through continuous changing gravity gradients causing constant energy losses. Would this same loss of intensity occur in gravity-free space if light traverses it at near infinite speeds? I do not think so since the Universe is young and we can see distant objects.

If the effective gravity in the universe is typically limited to gravity shells around stars and central galaxies, then the vast majority of our six- thousand year old universe is effectively gravity free and causes no impact on light and energy except for the proportional losses due to the inverse square distance rule.

Lensing of Light caused by Stars and Galaxies

Postulate L: Light from a background star or galaxy will refract producing an observable arc called lensing. Refracting light passing through the expanding gravity shell (bubble) of a foreground star will over time shift position or disappear due to the expanding shell as seen by an observer.

Starlight lensing may prove useful in establishing the effective limits of the interaction between gravity and light lensing around a foreground star. This boundary establishes a change in medium between an effective gravity field and ineffective zero-gravity. This boundary can be determined by measurements of the radius of lensing arcs produced by a properly positioned foreground star.

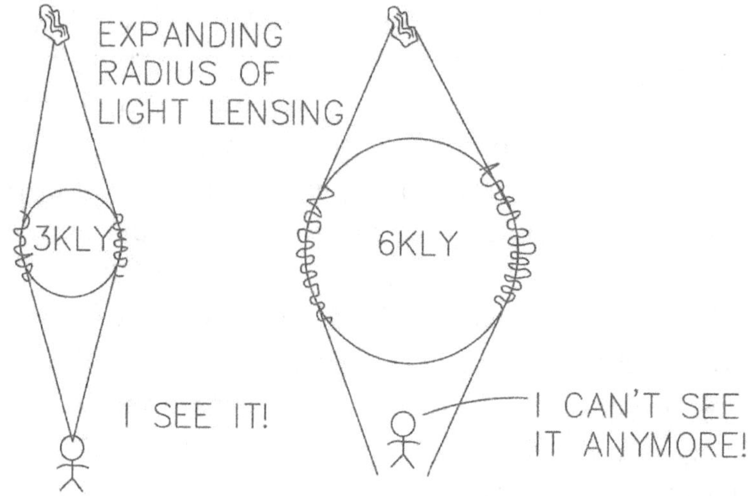

Figure 20: Measuring changes of radius of light lensing

Figure 20 shows how light refracts when passing through the edge of a gravity shell or bubble. The lensed arcs will also expand due to expansion of the gravity shell. In the above example, the arcs will double in radius as the bubble expands from 3 KLY out to 6 KLY.

We briefly touched on lensing of light earlier. In order for light to produce this lensing arc, it must pass immediately into and back out of a change of medium. It must also happen at the very edge of the change in medium in order to produce the lensing arc seen by a distant observer in the precise location to view it. This is important as it introduces an effective outer limit of the gravity bubble radius. Since the gravity shell (bubble) is expanding, then the image of the arc should over time, appear to make slight changes in its position and eventually disappear from an observer's position. Simple monitoring these arcs over time will prove this postulate.

We need to consider how light refracts producing the arc lens. Experimentation with light passing through transparent objects typically allows light to pass with little refraction accept at the outer edges or surfaces of the object. If this was not the limiting case, light would refract everywhere when passing through another field, we would not be able to see anything from space very clearly. We can draw a conclusion that light passing through the edge of a gravity field refracts producing an arc lens. Of course, the observer must be in the precise location in order to see this result.

Figure 21: A lensed arc produced by a background object

Lensing of background star light
https://www.space.com/41970-galaxy-cluster-gravity-lens-hubble-photo.html
NASA's Hubble Space Telescope photographed a stunning galaxy cluster whose gravity acts as a natural lens, bending and blurring the view of galaxies that lie behind it.

*The galaxy cluster, also known as SDSS J1138+2754, was photographed using Hubble's Wide Field Camera 3. When a massive object, such as a galaxy cluster or a black hole, falls between an observer (like the Hubble telescope) and a more distant target in the background, it creates a so-called gravitational lens that **magnifies** the observer's view.*

*Specifically, the powerful gravitational force of SDSS J1138+2754 **distorts the space around it, causing light from objects in the background to travel along curved paths.** As a result, the more-distant spiral and elliptical galaxies appear stretched into long, smudged arcs and scattered dashes, according to a statement from the Hubble Space Telescope. [Celestial Photos: Hubble Space Telescope's Latest Cosmic Views]*

I do not agree with the way this is presented since we know what caused light to bend, it is refraction of light passing through a change of medium.

Lets do a little review looking at this actual case of starlight lensing. The photo above indicates light from a background galaxy is bending (or curving) due light refraction caused by a foreground galaxy. We will use the term "object" for both stars and galaxies. In order for the light of the background object to refract as an arc, some conditions must be met. Just as observing light of rainbows, the observer must be in the precise location to view the lensing arcs of light. The light passing through the foreground object must catch or eclipse the background light at the precise angle. Finally, the foreground object's gravitation field will set the effective radius or position of the arc. If gravity shells (bubbles) exist, then over time we should be able to measure changes in the refracted arc as the gravity shell expands.

There is a reason an arc lens would not change its radius over time. We will later look at the ratio of gravity to energy (g/e) a little later.

Conclusion

In order for light to bend, it must react with gravity. Refraction is caused by a change in medium, the light passing from zero gravity into a gravity shell is a change of medium. This change of medium not only bends light, it will produce an arc at its effective boundary or radius.

If we look at light and energy traveling at the speed of light radiating in all directions, it gets pretty beat up as it is bent, refracted and dispersed. Gravity reacts with and disperses light from distant stars and galaxies and causes redshifts. If light were constantly influenced by gravity, its energy and intensity would undergo energy losses over great distances so much so it could not be detectable even by the Hubble telescope. Energy and light would benefit by not traveling great distances through gravity. The very absence of gravity in the vastness of space in a young Universe allows us to see these distant galaxies. If we look at light as energy traveling through space, while in zero gravity space, there are no losses due to refraction, dispersion and diffraction as there is nothing there to cause energy losses.

Energy reactions with gravity puts energy to work whether it is useful work or losses in efficiency due to it traveling through gravity. In order to see very distant objects, energy must necessarily be conserved since most of its journey is through gravity-free space. If we understand that light is not always traveling through fields of gravity, this would certainly throw a wrench into how we calculate distances and luminosities of stars. It would appear that more time light spends in gravity fields of other stars and galaxies, more is added to red-shift and that confuses it's actual distance measurements.

Throughout this chapter we challenged the curving of light to be caused by the principle of refraction of light causing it to take longer paths. Time dilation can only be a change of the distance light travels due to refraction since that takes light on a longer path requiring more time to get to its point of observation. This enforces the fact that gravity and light work together. Nothing is added in this chapter to obstruct light and energy traveling at unrestricted speeds through zero-gravity space outside of the bubbles.

Chapter 8

The Interactive Relationship Between Energy and Gravity

Postulate M: Gravity must possess an interactive relationship between itself and energy. Massless energy excited by gravity fields produces visible light and that responds to the design of the eye allowing it to see the energy as light. Without gravity, energy cannot be excited.

In this chapter, we are going go back to some basics in physics and in the process reinforce what we have covered so far.

On Earth, light and energy effectively appear as wavelengths across a wide spectrum of frequencies determined by its nuclear and chemical properties. Each property within the spectrum of energy reacts and responds to gravity at its specific frequency. Energy that is converted by gravity into visible light is said to be in the form of *photons* [30]. Photons react with our eyes in a way that allows us to see.

We learned earlier that gravity waves and light have the same speed. It cannot be by chance that two separate phenomena have the exact same properties with respect to speed. It is reasonable to consider that one reacts with and affects the other. From what has been discussed, it would appear that gravity is the agent causing light to transform into visible properties. Massless energy when in zero or weak gravity fields may not be able to convert or transform into energy vibrations producing photons. If this is the case, the Universe must be filled with undetectable energy.

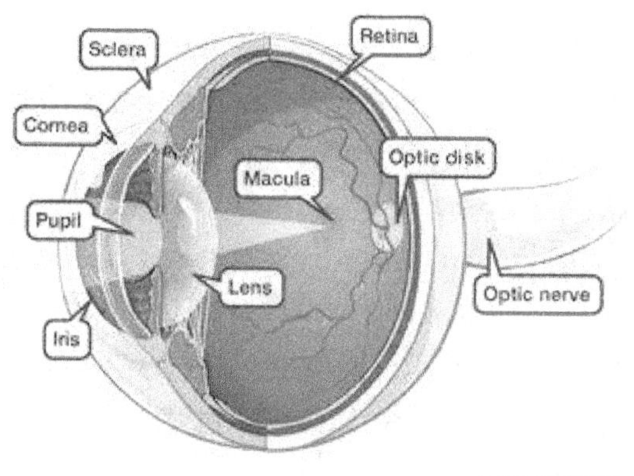

Figure 22: The irreducible design of the human eye

This postulate has deep implications concerning our ability to see, as we are betting on the chicken or the egg coming first. If the eye design came first, it follows that the universe must have been by *external* design. If evolution is the case, it must also include the evolution of "irreducible design"[36] of the human eye which requires five independent systems to have all evolved together in order for us to see. This relationship elevates any chances for *evolution* from the first order of extreme unlikely chances up to the fifth order.

Naturalism which includes evolution and the Big Bang places the chances of our very existence *inside* of this universe. The speculation necessary to believe *naturalism* introduces "irreducible design" on top of the already enormous challenges for evolution. Since design is apparent everywhere in both the physical and biological Universe, perhaps it is easier to accept that our existence and our ability to see is due to an *external* influence or Creator. Its like the old adage of a billion monkeys banging on typewriters and coming up with the Lord's Prayer. It aint gonna happen. Besides, how many quantum universes would have failed before we landed on this one?

Dark Matter's Relationship with Gravity

> *Ref: "Creation Magazine" Creation 40(4) 2018 p19 (Sarfati):* All matter in the Universe seems to be built up of 17 named fundamental particles that interact by four fundamental forces. The particles and all forces **apart from gravity** have been described as the **"Standard model"**

What is interesting about this *"Standard model"*? Dark Matter particles are generally described as anti-matter. If Dark Matter is made up of anti-matter particles, it is generally accepted that matter and anti-matter when combined react violently releasing energy. Theoretical physicist can offer much more information on this subject but since the existence of Dark Matter is still speculation, we will not get into that within this work. Dark matter remains an undiscovered mystery.

Gravity and the Four Fundamental Forces of Matter

Referring to the above *Standard model,* this is simply enforcing the concept that without the physical material we call matter, gravity cannot exist since the fundamental forces of matter create gravity. But what about energy existing outside of gravity? Gravity forces produce energy from nuclear fusion in stars. This fusion energy is not part and parcel of the four fundamental forces contained in matter. Therefore energy produced in gravity becomes an independent product and can extend and exist outside of gravity.

Again, our concern here is with the comment above about forces *apart from gravity* describing all known forces except gravity. It is reasonable to claim that all described fundamental forces contained in matter actually *create* gravity. If gravity forces are the cause of nuclear fusion within matter, this newly formed energy becomes independent of matter and gravity so therefore can exist outside of the gravity bubbles. Paradox!

In the case of energy produced within nuclear reactions, from the moment of its creation, it can exist in two states. It can travel with gravity at the speed of light but as the distances increase, gravity weakens and its intensity dissipates. Energy also radiates instantly expanding throughout the Universe but also dissipates due to the inverse square of the distance rule.

The Film Maker

Let's consider some analogies in order to understand how something may have come from nothing. A film maker creates and produces a comprehensive movie about the universe and everything it contains. To date, no one has seen this movie so a grand premier is held for all to see. The very instant the projector shoots the images onto the giant screen, the viewers are in awe of the wonderful works of the filmmaker. An enormous amount of time and thought went into the masterpiece but for the viewer, nothing was known about it until the moment it hit the big screen.

Knowing what it takes to produce a great movie, we might consider that the design and creation of this movie was the work of an amazing film-maker. The time it took to make this film and the tedious work was done, so all that was needed was the flipping of the switch for all to see. This analogy does two things, it removes the "magic wand" concept of creation and it necessarily removes the completely flawed concept of a cosmic explosion creating an irreducibly designed eye. Most likely, everything in this world and universe was thought out in great detail "before" our creator flipped the switch on the projector.

Consequently, a random big bang lacks forethought and design and incorporates chance; how many times and chances would have had to take place before a single galaxy would have formed? The answer is impossible to know and it pushes time required for naturalistic beginnings of life far over the horizon.

By incorporating logic based upon all human understanding, our beginnings must have had some help! As the famous Sergeant Jack Friday said "just the facts ma'am". Well the facts are stacked against the Big Bang so let's move on with our postulates to something else with more solid-like science supporting the documented and factual history of the Creator!

All light is a product of energy being generated or produced somewhere for example in a star. For now, let's limit our thinking to visible light that we can see with our own eyes. It seems as previously discussed, that light is a product of, or a derivative of energy that is transformed into some useful purpose which obviously allows us the ability to see. We can reasonably claim gravity reacts with energy producing visible photons as they are converted from massless energy into photons of light.

Einstein's Relativity and Gravity-free Space

For the moment, let's place ourselves outside the bubbles in the Universe where there is no gravity forces of any kind. Using Einstein's $E=MC^2$, lets substitute Mass with "weight divided by gravity" (w/g) since (M = w/g) so we get: $E = (w/g)C^2$. We can conclude that energy (E) in gravity-free space could not be calculated since as gravity approaches zero, the equation becomes undefined $(E = (w/0)*(C^2))$ (cannot divide by zero).

This does not mean that energy is not there because we know that energy and light traverses the Universe through vast voids of "ether". Since we do not know what "ether" would be, we will simply call it micro or zero-gravity space. A reasonable conclusion is that energy can also be independent of gravity and therefore $E=MC^2$ is not valid unless gravity is present. This begs the question: how much gravity is needed to excite energy in order for it to be visible?

What happens to Newton's Law's in zero-gravity?

Newton's second law of motion describes the relationship between an object's mass and the amount of force needed to accelerate it. Newton's second law is often stated as (F = ma), which means the force (F) acting on an object is equal to the mass (m) of an object times its acceleration (a). As a reminder, anything multiplied by zero equals zero and anything divided by zero violates math rules and is undefined or not possible. Newtonian physics deals with gravity, matter, energy, time and accelerations. Let's see if we can uncover anything else about Newton's physics we can compare to energy and gravity.

When we speak of and perform calculations using mass (m), what are we really doing? We are creating a "ratio" of an objects weight divided by gravitational attraction (m = w/g). Mass is simply a ratio. This is very useful and meaningful for calculating forces on or around the earth's and solar system's gravity fields. However, as we venture outward toward empty space and away from any meaningful or strong gravity attraction forces, mass ratios begin to lose it's meaning. Weight of an object is considered "fixed" because every material object regardless of where it is in the universe still has "weight" which is properly called "inertial mass". For example, it would take more energy to throw a lead ball than it would take to throw a baseball of the same size. So the lead ball has more weight or inertial mass than a baseball. If there is no gravity, mass equals weight divided by zero. This is meaningless but objects in zero gravity still have inertial mass. The point is that pure energy has no mass or weight, so none of the above applies.

By transposing (m = w/g) into (w = m x g), as g approaches zero, as it would in deep space, there would be no external forces acting on any object of weight whether it was stationary or in motion. We cannot change the weight (inertial mass) of an object, but if we replace the object with pure energy which has no mass, how would energy itself act differently. If weight is zero such as with energy, then what would be the state of energy in a gravity-free environment? Newton's laws are not applicable without gravity or some force acting on it. We need something else. We are building the case for what happens to energy in zero gravity space.

Conclusion

In conclusion, we are saying that energy does not obey Newton's laws or Einstein's Theory of Relativity in zero or micro-gravity space. If that is the case and of course it is subject to scientific verification. Then it appears light and energy can transcend gravity-free space with no limits based upon Newton's Second Law of physics which includes gravity. We will discuss more on this later.

Chapter 9

Energy and Light from Nuclear Reactions

This chapter will provide more information and greater detail on some materials already presented while introducing new ways to look at energy and light. We typically think of gravity as a steady-state force pulling us to earth. It is important to think of expanding gravity fields in terms of waves traveling at the speed of light. Gravity radiates outward and its strength decreases in proportion to the inverse square rule just as light.

We can conclude that the so-called *solar wind* is the motion of gravity carrying with it sub-atomic particles left over from the sun's nuclear reactions that can damage earth if it were not for our magnetic field to protection us.

Postulate N: Strong gravitational forces must be present to produce Nuclear Reactions inside a star. Within a star's nuclear fusion reaction, each atom's reaction is independent of every other reaction occurring at billions of inharmonious times. Without gravity, energy cannot be created or organized for useful purposes. Massless energy escaping gravity cannot be expected to respond the same way it does in gravity.

This relationship begs the question of what happens to energy and light if it escapes gravity. So can pure massless energy and light exist in some state of electro-magnetic radiation or some other *undetectable* form that does not react to gravity? To examine this condition, we must perform some type of test that shows how massless energy radiation can escape gravity and can travel at speeds greater than gravity and light.

Since we cannot ourselves escape gravity, we must determine a way to test this postulate for light speed between two independent gravity fields. There is no direct way to do this but creative science can use other related methods already in use to help with this test. For example, we have shown a way using light lensing to measure the expanding radius of a star's gravity shell with respect to our six-thousand year-old Universe. I am sure there are creative ways to use this information. Faster light travel will establish that the age of the universe is independant of the speed of light. It is reasonable to seek answers for understanding that energy and light must have escaped gravity to travel the vast distances of empty space.

To be sure this relationship between energy and gravity is true, we consider that the same thing happens within a chemical reaction such as in flashlight batteries producing heat across the filament of a light bulb and that heat is immediately organized into wave forms of both heat and visible light. Even a flashlight has mass, it would be interesting to know if a flashlight could be transported out into zero gravity and turned on, would we be able to see its light?

Postulate O: The First State of Pure Massless Energy: Energy and Light produced inside a nuclear reaction can exist in two independent states: The first state is pure massless energy in some form of radiation that does not respond to gravity and has instantaneous velocity. This state of existence is supported by the 2nd Postulate of Relativity which states "The speed of light in a vacuum is the same for all observers, regardless of their motion relative to the source." Therefore, light is always traveling at the speed of light even if the observer is also traveling at the speed of light. This state of energy travels at infinite velocity and will extend out in proportion to the inverse square law[31] .

Postulate P: The Second State of Pure Massless Energy: This energy is the form that is converted into useful work by gravity. The second state of energy obeys the laws of physics and relativity. In this state, light is converted into photons and also obeys the Inverse Square Law.

Since light is observed by all observers traveling at the speed of light, then light must accelerate ahead of the gravity field removing itself from the effects of gravity which is the producer of vibrations or wave lengths. That light would exist throughout the Universe at the same time subject to the *Inverse Square Law.*

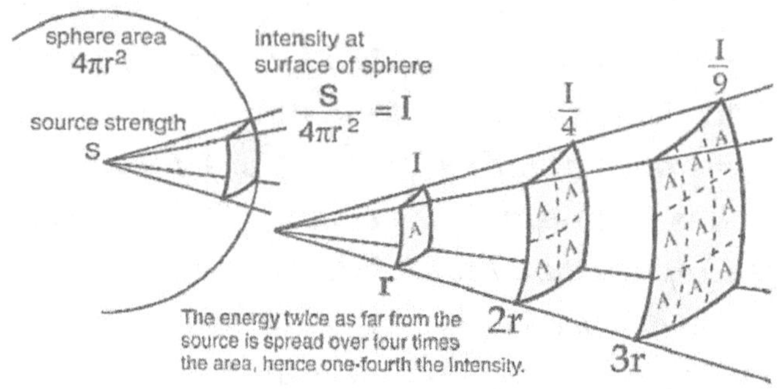

Figure 23: Inverse Square Law

Figure 23 explains that light and energy are proportionally reduced by distances from the source. Light distributed over a nearby surface loses intensity as the distance from the source is increased since the same amount of light has to distribute over a much larger area. This law holds whether in gravity or zero-gravity.

Consider that a star is created and it instantly comes into existence. At the very moment of creation when (t=0), nuclear fusion reactions begin and the star's gravity field begins to expand at the speed of light. If both occur at (t=0), energy is radiated outward at a velocity greater than the speed of gravity; it is undetectable since it radiated ahead of gravity.

We postulated this energy cannot be detected outside of gravity. The energy radiation is traveling at speeds near infinity since there is no interaction or anything to stop it. That energy will continue radiating outward throughout the universe. As it radiates outward, its "potential" energy is reduced by the inverse square of the distance just as with gravity. Meanwhile, back at (t=0) the gravity field begins its expansion and will continue to expand indefinitely but is also reduced in strength by the same inverse square of the distance principle.

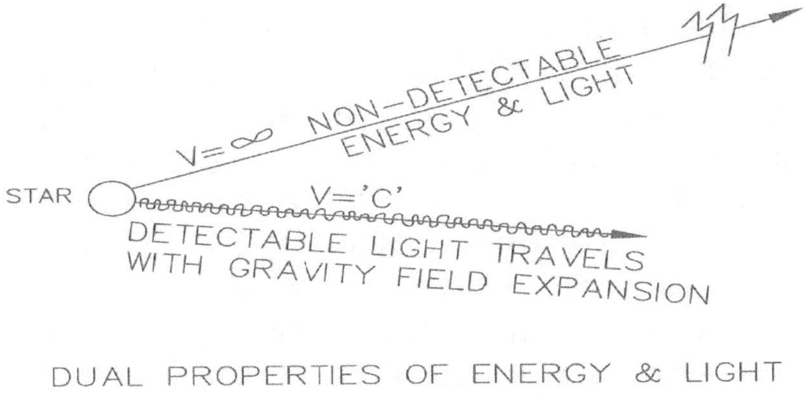

Figure 24: Showing the two states of energy and light motion

Pure Energy does not contain mass

It is well accepted that when matter is subjected to certain nuclear reactions, parts of it are converted into energy. Energy then, is absent of matter even though it could possibly be contaminated with bits of matter including quantum (sub-atomic) particles. These particles travel with expanding gravity fields making up the Sun's solar wind.

This presents some interesting speculation as to what happens when energy is accelerated to the speed of light. Matter itself cannot reach that speed as Einstein predicts. If pure energy contains no mass, as it is created, it must accelerate to the speed of light instantly as it is carried with the speed of gravity. This is explained by Postulate L , *"The Second State of Energy and Light"* above.

Consider what could potentially happen if energy produced in gravity extends into an extremely low gravity environment. From previous chapters, if energy (or light) contains no mass it would not be subject to Newtonian physics or Special Relativity; only two possibilities are available: first, is that energy could not exist "outside" of gravity, or second, it can exist outside of gravity but not subjected to the laws of physics or relativity. Where can we go that is outside the effects of gravity? We have answered that assuming the existence of six-thousand light-year expanding gravity bubbles. It is anywhere outside the bubbles encompassing nearly ninety-five percent of the Universe. The universe is estimated to be only about five percent ordinary matter.

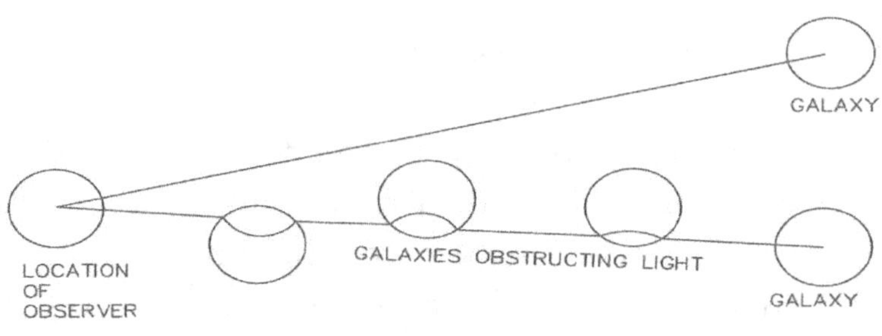

Figure 19b: Gravity fields bending light and causing red-shifts

Lets again look at a light source that has traveled through several gravitational fields. As we look deeper and deeper into space, of course we know that light loses energy and efficiency while in gravity. If we isolate a particular point of light from deep space and study it under this theory, we must consider how many other gravity fields it penetrated and then we add the distance through our own gravity field. Light would only lose energy efficiency when passing through gravitational fields. This is to say that we cannot truly determine the intensity of a light souce if it traveled through multiple gravity fields of unknown numbers and distances regardless of the age of the Universe.

If this is the case, when we look into deep space and attempt to compare the intensity of two light sources, we must be careful as we have no idea of effects of gravity when compared to any other source of light. It would be very difficult to determine accurately which source was actually nearer based upon red-shifts alone. We would only conclude that the weaker the source or greater the red shift, the more time it has spent in the effects of gravity.

A logical arguement opposing this is that the Universe is old and gravity is present everywhere. If that is the case, when you compare two galaxies as in the figure above, how would you sum up all of the times light bent and refracted due to ever changing gravity density and gradients as they are impossible to summarize. We know the Universe is huge. We do not know the amounts of gravity light has traveled through.

Lets look at the definitions of pure energy which includes light. When mass is converted such as in a nuclear reaction and becomes pure energy, it no longer has mass. Einstein's famous equation $(E = mC^2)$ states that the amount of energy produced is its mass times the speed of light squared. Mass in that equation is defined as $(m = w/g)$ which is the ratio of weight divided by the gravitational attraction at any particular place in the Universe within a gravity bubble. Note that as gravity (*g*) approaches zero but not zero, the amount of energy produced at the speed of light becomes incalculable. Commonly stated, if an object of mass is accelerated, it would require an infinite amount of energy to reach the speed of light.

By definition, energy produces work when it transfers from one body to another or from one state to another. We find ourselves interested in this process of transfer in certain conditions. Our ability to see is a result of energy in the form of light being transferred to our eyes giving us the ability to see. We must also consider the role gravity plays in this process as well.

Pure energy emittance

Light can be produced in many ways including simple chemical reactions using batteries and filaments that heats and glows as electrons pass through it. Light is a part of energy that is produced from both nuclear electrical and chemical reactions. The mystery we are unfolding is how is it that energy produced by any useful method will reveal itself in the form of waves?

All stars are sources of nuclear energy and contain specific spectral patterns corresponding to the material the star is composed of. Light in wave form allows us to examine a star's composition using various instruments including spectrometers. What else can produce light differently from light in nuclear reactions of stars? Would a chemical reaction producing light organize into wave-forms without gravity? To define this, we needed to know the definition of energy, how it is made and the spectral analysis of it. A star's total energy emittance occurs over a very wide range. Our focus is on light emittance.

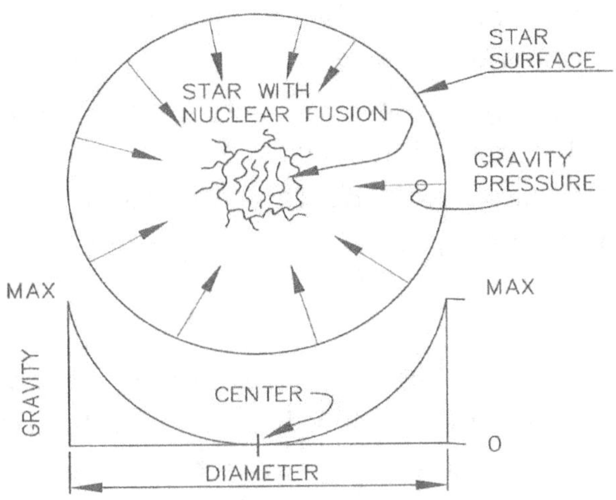

Figure 26: Gravity pressure is greatest at the radius of a star

Gravity harmonics and energy:

Figure 26, shows random nuclear chain reactions within a star. Note that in the center of the star the gravity force is near zero in the midst of the fusion. Perhaps it is the gravity intensity at the radius that organizes internal energy and light into wave forms and accelerates them to the speed of light.

How can nuclear energy be produced in wave form if billions of independent chain reaction events are raging at different times separated by millionths or billionths of seconds? What force would be necessary to "train" or "synchronize" them to respond together producing wave forms? Theoretically, fusion does not produce a single detectable wave or range of waves because it produces a massive conglomeration of inseparable pips or spikes best called "noise".

A micro-instrument inside a nuclear reaction would not see any frequencies but a massive noise-state of energy. This begs the question, how is energy actually generated in wave form? Nuclear energy *noise* when created must exist in some state or form other than vibrations. What is it in the universe besides gravity that would synchronize energy into well organized motion that holds to the principle of waves and harmonics?

We may draw a conclusion that gravity is the only force available that could accomplish this alignment. Energy and light must necessarily be strongly influenced by gravity. It further postulates that if energy could escape the forces of gravity, it could no longer be measured as waves, in fact it may not be measurable. This leads us to consider whether pure energy can escape the effects of gravity and if so under what conditions? If it could, then what would constrain it or govern it?

As we discussed, there would be no ranges of harmonics, frequencies or wavelengths inside nuclear reactions since each reaction occurs at its own independent time. When energy is produced, as it is in gravity, its speed instantly accelerates to the speed of light which is the speed of gravity. Energy literally surfs on gravity waves. As energy exits the nuclear reaction zone inside a star, gravity causes it to get its act together by forming it into measurable waves.

It is generally accepted that matter is "partially converted" into energy within a nuclear chain or fusion reaction. We know that the process is not one-hundred percent efficient as there are vast amounts of atomic particles left in residue. These particles are measurable at velocities all of which are less than the speed of light. So we ask, what is the velocity of pure energy as it radiates out from the nuclear reaction? We stated earlier that light is a product of energy within the spectrum of our ability to see with our eyes or with various types of instrumentation. All energy we are able to measure regardless of whether it is visible within our eyesight or with instrumentation is in wave form and at broad ranges of frequencies.

Looking again at the phenomena of energy as always measurable in waves or harmonics. Take a look at say one particle of matter being converted into energy in a nuclear fusion reaction. As the particle is physically changed, the energy produced began a vibration or harmonic that we measure as waves over a broad spectrum. However, in a reaction where billions of particles are being converted into energy within a very small period of time, they are not all at the same time and could not be harmonic. How would they become harmonic so that they are all vibrating or producing measurable waves? It is reasonable to think that without the influence of gravity, all of the billions of energy bits could not arrange themselves into waves. How would we be able to measure light wavelengths due to so many billions of individual reactions washing each other out?

If energy escapes gravity and is no longer detectable, then energy that is out there in zero gravity can do nothing except travel very quickly across the Universe until it reaches some other gravity field.

Lets be careful here so that we are not kicking the can of reason a little further down the road by claiming that energy cannot exist outside of gravity, the claim is it cannot be measured or seen outside of Gravity. We cannot assume for a moment that energy produced in a gravity field is always measurable in wave form when it escapes gravity. If that is so, could we draw a conclusion that energy that has escaped gravity cannot be measured or put to work and remains in its purest form we called *noise*.

Postulate Q: Energy traverses zero-gravity and possibly very weak gravity fields in empty space at infinite speed.

Lets now look at a theoretical experiment. We place a nuclear bomb at a vast distance from any object in space where gravity is virtually non-existent except for the gravity of the device itself. We set the device off into a massive explosion so that all material contained in the device is converted into energy. This reaction took place in a gravity-free environment to the extent that all of its mass is converted to energy and becomes independent of gravity.

Of course the energy is dispersed into empty space in all directions. Immediately following the blast, there is no gravity for the energy to react with so it is not measurable in wave form. There is nothing to constrain the energy velocity since it is not in a gravity bubble and does not have mass. Special relativity cannot apply to this test as it lacks gravity or wave forms. There are no rules, as this energy radiates at instant velocity into empty space. There was nothing to constrain its velocity, so it could exist at all points in zero-gravity space at the same time which was basically (t = zero). Of course the inverse square law still applies so it will weaken.

We now start with the examination of light with what we know and what we do not know about energy and light. We know that if gravity affects light and it bends, it must have mass, but we also know that according to Einstein, nothing with mass can approach the speed of light. We start this phase of the process with a sense of confusion. But we do know the speed of light since we have measured it. We must consider the location of our laboratory and the conditions used to establish these measurements and theories.

Of course we live on earth and earth is influenced by gravity and we know that gravity influences light waves so we should conclude that what we know about light is only within the presence of gravity. We are very limited.

For the moment, consider if there would be a difference placing our lab in a gravity-free environment. What could we learn that is different than what we know from our earth based labs. Lets ask ourselves is it possible that light is a completely different phenomena outside of gravity? As the postulate suggests, is it possible that gravity is the acting force that creates those wavelengths at the speed we are so familiar with? There is no other reasonable explanation.

In truth, we do not really know if we can measure energy and light outside the effects of gravity, so now let's think outside the box. Lets start by thinking what would light be like outside of gravity. If gravity is the necessary force that causes forms of energy to oscillate (vibrate) at all these different wavelength frequencies, then outside of gravity there would be no oscillations and consequently there would be no visible light for our eyes to see.

Consider that visible light can only be produced in a field of gravity as all massive bodies create measurable gravity. Now lets propose that the produced energy leaves the gravitational influence where it was created and travels to another gravitational environment where it re-appears. If we are located at the second environment and we can measure it, what would we find? We can only prove that the velocity is in agreement with what we know, but we know nothing of what happened to it as it traveled across the bubble-free zone. We cannot measure the distance it traveled.

Lets say we are great distances away from each other. I am positioned at the edge of my gravity bubble and you send a beam of light from the edge of your gravity bubble. The beam travels only outside of the bubbles and is not affected by our gravity environments. If this postulate is correct, the beam is immeasurable and invisible in between these two bubbles although it still exists in a state not influenced by gravity. There would be proportional losses of the light due to the inverse square rule.

It would not be a giant step to conclude if it was not affected by gravity, it would take no time to travel that non-gravitational distance. It is simply energy radiation instantaneously distributed throughout the Universe. Einstein's Theory of Relativity only applies in Gravity.

Postulate R: Gravity is the producer of wave-form energy. Gravity must necessarily transform energy into detectable forms of energy . The state of massless, pure energy itself does not exist in or possess wave forms. An as yet undefined phenomena causes gravity to react with energy producing detectable energy.

Universally speaking how do we know if each and every "gravity producing" star or object in the universe is in wave phase with each other or whether it is even necessary? It does not seem possible to directly measure the intrinsic internal properties of gravity as a producer of vibrations. If intrinsic gravity operates across an extremely wide range of internal frequencies, then all energy products such as light and heat may operate at some fractional or proportional or ratios to gravity causing them to vibrate. We are so limited, since we only know about the speed of gravity by observing gravity induced waves and they travel at the speed of light.

Could there be some universal frequency coordination so light waves are universally in sync with each other? Since it is known that gravity is defined as an attractive force between two objects, what would happen if two star's produced light waves that were exactly one-hundred eighty degrees out of phase with each other? Would each of the star's gravity also repel each other or would their light just cancel each other out? We must beg the question of whether gravity is universally coordinated or whether each independent gravity field operates at its own frequency. Presently, there is no way to know.

If "out of phase" gravity fields exist, that could explain the repulsion and expansion of the Universe.

If gravity forces operate in wide ranges of wave form, gravity waves in effect would produce pulses measured as waves but most likely half-waves since the gravity force must always be positive. Then are gravitational forces only pulsating at half-wave frequencies? That would mean we could measure gravity forces as pulsating and not as a continuous steady state force.

If gravity waves between objects are not in sync, could that mean that at instances when mutual waves are out of sync there would be zero attractive forces working toward each other? These are all questions we have no answers for. We know that light waves are full cycle waves and if gravity waves are also full cycle waves, then the waves when plotted must always remain in the positive to zero range. In other words, gravity waves could not go positive then negative. If they did, perhaps we could invent anti-gravity propulsion.

Conclusion

Considering when pure energy is being produced in a nuclear reaction, it is transformed from random noise into harmonics in a variety of wave lengths . What universal force would do this? There are no other forces in the universe that we know about except gravity that could accomplish this. We could draw a conclusion from this information that if it took gravity to do the aligning, then when energy escapes gravity, it would exist in its original state and would not be seen or measurable in wave lengths. There would be nothing to constrain it. Then pure energy could radiate the universe at infinite speed until it once more contacted gravity fields.

Chapter 10

Light Superposition

Superposition of Light from Two independent Gravity Fields

This chapter is a bit of a carry-over from the last chapter. This is admittedly aimed more at physicist for purposed of research to determine the independence of gravity fields. An informed reader with basic physics should be able to follow.

Physicists are familiar with spectroscopy and use oscilloscopes and equipment to observe and measure a variety of light frequencies. This section proposes a very difficult experiment calling for light from pairs of stars at various locations to be superpositioned onto a test surface in a lab to analyze and compare test results.

It is unlikely this experiment would produce any useful results as there are so many variables. Results could be in agreement or disagreement with the overall postulates presented in this work.

> *Astronomers* can also measure motions on the Sun and *stars* by measuring changes in the wavelengths of emission lines, or by the shapes of emission lines in the spectra. Motions can be measured because of the Doppler effect, which changes the wavelength of sound waves or light waves from a moving source.
> sunearthday.nasa.gov › materials › spectrabook_ver2

> A *stellar spectrum* can reveal many properties of stars, such as their chemical composition, temperature, density, mass, distance, luminosity, and relative motion using Doppler shift measurements. en.wikipedia.org › wiki › Astronomical_spectroscopy Astronomical spectroscopy - Wikipedia

Observation of light waves requires very sensitive and accurate instrumentation. Light frequencies are so high the human eye cannot detect them. Motion pictures change picture frames just twenty-four times a second and we cannot see that. This chapter calls for experimentation to determine if lab testing using superposition will yield any further information about light waves produced in different gravity fields farther than our own gravity field.

Just like waves in the ocean, light superposition occurs when two light wave sources are combined so their light intensity either amplifies or darkens when being observed. An observer should be able to measure superposition of stars varying in intensities over time. While this is not normally seen by the human eye due to high frequencies of the light sources, these differences may be measured accurately in labs. It is possible that distant star light from two combined sources may fade in and out under superposition conditions.

This experiment may also yield results in determining if the light of "variable stars" is the result of unintentional superpositioning rather than the star itself changing luminosity.

In this difficult lab experiment, we test light from two separate gravity fields to determine if the gravity fields themselves are in phase with each other. This could say something about whether gravity sources are in phase throughout the Universe. This test must consider relative motions of the stars and gravity fields. While this test may be virtually impossible, it begs the question: can two separate gravity fields produce independent, out of phase sets of light waves based solely on their gravity frequency which is isolated from each other?

Postulate S: Independent out of phase gravity fields may cause light interference between objects in deep space and may provide additional information using the principle of "superposition" which states "When two of more waves of the same nature travel past a given point at the same time, the amplitude at the point is the sum of the instantaneous amplitudes of the individual waves."

Theoretically, light produced in two separate gravity fields are wave independent of each other, unless there is some other universal force synchronizing them. Separate gravity sources may or may not be in phase synchronization. However, light that traverses zero-gravity space may lose its wave properties altogether returning to pure energy in "noise" form.

If a light wave is stretched due to infinite velocity, its wave signature would be flat, not producing visible light waves. Then as it reaches our gravity shell where it once again compresses back to the speed of light it is restored into energy waves. This calls for experiments into how sources of light may or may not synchronize at a test station where the two sources of light are combined.

This experiment may provide information about light transitioning into zero-gravity space and no longer acting as waves but flatning out to act as massless instantaneous velocity vectors, either in potential or kinetic energy form until it reaches our gravity influence.

A lab test would require initially synchronizing two sources of light from two stars greater than 12 KLY distance, then observing any frequency or wave shifts as the earth rotates around the sun. If the star lights stayed in sync, it may indicate synchronization is a local gravity event when the light comes in contact with our 6 KLY gravity boundary.

If energy could be detected in gravity free space in wave forms, then all of the energy sources may not be in harmonic synchronization with each other. This is similar to ocean waves coming from different wind sources and producing a much greater wave, or simply canceling each other out.

Although light frequencies are very high, and its wavelengths are very short, if light from different sources did not synchronize, then we should see all types of blinking light anomalies when we combine them for observation. Light physics behaves just like waves on the ocean and can either combine in constructive amplitudes or negative amplitudes.

This postulate is introduced to give thought to what may be further understood about light passing through gravity-free space.

If we observe light from two stars and superimpose them, what should we expect? If the stars are in motion relative to our lab, the rate of change of superposition would allow us to calculate velocity changes in relative motions.

Conclusion

These types of tests may be very difficult to perform but is reasonable and possible. It may prove that light from each star has its own wave phase but as it left their respective gravity fields their waves diminished and remained flat until the time they entered our gravity field. This would imply that light waves are local gravity induced events having nothing to do with the distance of their source or relative motions or time spent traveling through gravity-free space. Admittedly this chapter presents some very challenging tests for physicist and astronomers.

Chapter 11

Examples and Explanations

What in the physical Universe can exceed the speed of light?

In previous chapters we discussed the phenomena of energy when it escapes gravity as being undefined in any terms or formulas, including Newton's and Einsteins. There are no laws or rules to constrain its speed. In order for energy to do anything, it must be within the presence of gravity. So energy can escape gravity and when it does, we cannot see it or measure its speed. Like theories of Dark Matter, it becomes an *ether* filling space in vector-like forms directed toward some purpose for use or observation.

Coulomb Field Speed velocity determination

The study of physics includes the study of electro-magnetic fields. Coulomb field effects may act the same way light energy does in that the field speed seems to be incalculable.

Is there anything else in the physical world that can exceed the speed of light? Referring to *Figure 27,* do electric and magnetic fields travel instantaneously with electron charge along a wire or do they lag? The fields traveling with the electron charge are either created instantaneously or could be lagging the motion. However, the fields may be considered an integral part of the charge itself meaning the fields are not separable from the energy of the electron. This phenomena is introduced for purposes of thought regarding the possibility of instantaneous velocity of the electric and magnetic fields. There are differences of opinions regarding the speed of electric and magnetic fields.

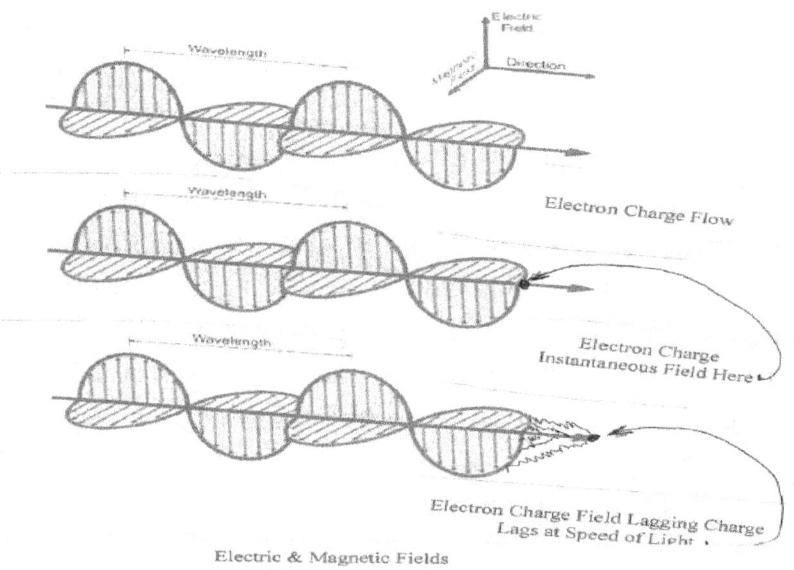

https://physics.stackexchange.com/questions/461766/comprehensive-nature-of-light-transverse-vs-longitudinal-waves-and-ray-optics

Figure 27: Are electric and magnetic fields created instantly or lag the flow of electron charge?

Electric current flow in a wire is a series of energy charges transferring through electrons along a wire. It is not likely the electrons themselves are moving since they have mass and cannot move at the speed of light. The movement is limited by the speed of light traveling along a wire. Coulomb field speed is introduced here only to consider what is taking place when a wire is charged and the charge is moving at the speed of light along the wire producing both electric and magnetic fields.

Can the speed of light also accelerate in micro-gravity space?

How do we know what happens to light as it traverses the universe? We know that nuclear fusion energy is created in very high gravity, and instantly accelerates to the speed of light and rides as if attached to gravity waves. This active light which is detectable will continue out to a distance where gravity becomes so weak that energy can no longer be "activated or excited" by gravity. It exits the bubble then perhaps its wave form flattens out so it is undetectable. The process of flattening out may be the boost that accelerates it to speeds near infinity. When it contacts a new gravity bubble, it compresses and slows down back to the speed of light and its wave form is restored. This concept of light travel and speed adds to the mystery of what happens to light under varying conditions.

Can we determine a ratio between gravity potential and energy potential working together to produce visible light? Consider (g/e) as the ratio of gravity to energy. Without getting into the physics of the actual potentials, let's assume that if the ratio falls below "1", that is, the gravity potential is weaker than the energy potential, then below this point, energy flattens out and cannot do work or be detected.

We know mathematically that gravity between massive objects like the sun and the earth remain pretty strong, but what about gravity on a 160 lb man in deep space? Gravity can become very weak. What about gravity on massless energy called light? There is no gravity attraction in the first place except when converted into photons. So basically, (g/e) ratio is zero when there is no gravity, so energy and light are not affected or constrained.

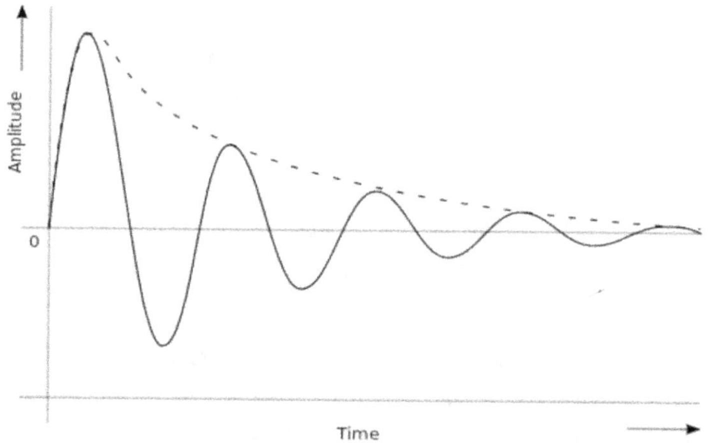

Figure 28: https://en.wikipedia.org/wiki/Damped_sine_wave

Refering to *Figure 28*, this sine wave may represent the relationship of light waves flattening out and accelerating when the (g/e) ratio falls below the value "one". If it falls below "one" but micro-gravity still is present, then what happens to the speed of light?

Our study of the speed of light is a neighborhood science. Consider that by definition, the speed of light is the speed of which useful energy is riding on gravity waves. We do not know what happens to it in deep space and whether or not it would still exist in wavelength form especially if there is no gravity potential strong enough to produce the wavelength.

There are two distinct possibilities, (1) If light can only exist within the effects of gravity, then it cannot cross micro-gravity or zero-gravity space. We could not see light from distant galaxies. (2) Light can traverse deep space unconstrained by the speed of light as an unknown property that is not the same as light under the influence of gravity. This light energy may not be bound by the speed of light. Later we will look at a test for this condition.

Postulate T: Gravity affects all forms of energy in ways that can bend it which means it must have some form of mass and this mass is then burned off or dissipated as it refracts and disperses while traveling in gravity.

So to the contrary, we could say that all energy and light existing outside of gravity is everywhere and at all times; it is only gravity that reacts with light giving it properties we can see and measure. If we can prove this, then all constraints on the age of the universe are removed since the speed of light can only be measured in gravity, and cannot be measured in deep space outside the bubbles.

So for now, looking at our universe, we see light coming from every direction. It is only slowed down or given that appearance within the effects of gravity.

Can a Spaecship Escape Gravity and Speed Through Space?

With God, all things are possible. Lets put our imagination to use in an interesting way. "Startrek" was and is a fascinating TV series about the wonders of the universe. Of course Captian Kirk's ventures encountered all sorts of life throughout his travels. If we apply our minds to the shear size of the Universe and the size of each galaxy, wouldn't our own Milky Way Galaxy be enough room to roam if it were physically possible? If we travel to the limits of a gravity field, our space ship still has inertial mass (weight) and that takes energy to accelerate. Even if we disregard Einstein's Relativity, it takes an enormous amount of energy to accelerate a space ship to anywhere near the speed of light, then when we get to wherever we are going, we have to slow back down and that takes the same energy as it did to accelerate. Also, there is a huge risk of attempting to pass through our own Ort Cloud!

Let's consider the Milky Way as our playground since it alone could contains millions of stars with possibilities for many planets similar to earth capable of supporting life as we know it.

The Judeo-Christian God has identified us as a "Creation in His likeness". Then by his will, if he created life on other worlds, they would also be in his likeness. Since the Universe is so large, if God did create life on other worlds, I am reasonably sure he would have moved them from beyond our reach. If we knew there was life similar to us within the Andromeda Galaxy, how would we ever get there? Even if God has given us a future and ability to explore his Creation, by stretching physics and time as far as possible, only a very small portion of the Milky Way would be our playground.

The actual physics of space travel is real and challenging. Since any type of spaceship built by mankind would have a significant amount of *inertial mass* and that takes huge amounts of energy to accelerate . We do not have an energy source that could reasonably propel our starship to an effective speed to reach our closest star.

Let's say we did have an energy source, could we reach the speed of light? That of course is debatable but I think if we had sufficient energy to propel our ship, we could reach any speed regardless of restrictions for the speed of light described by Relativity. That's just plain physics. We would still have to control the starship and stop at our destination.

Conclusion

There is a fundamental problem for us reaching out to other worlds. We are damaged goods. Mankind has sinned against God. This is common knowledge, if as a reader, you do not know this, you are missing very important information about the destiny of mankind. When left to our own devices, we corrupt things and we break down relationships intended between ourselves and our Creator. If we could reach a perfect world out there, we would destroy it with our sinful nature. We simply are not responsible enough to conduct our own affairs. When we ignore this wisdom about mankind, we are the problem. When we make our own rules and "allow" sinful, evil people to govern our affairs, we are left without hope of anything else. Is this the kind of relationship we want with each other – or our Creator? It would not be wise for us to contaminate other potential worlds. If you do not believe mankind is destructive, please brush up on our history.

New King James Version
By faith we understand that the **worlds** *were framed by the word of God, so that the things which are seen were not made of things which are visible. Hebrews 11:3* This verse is a clue for an explanation seen in later chapters.

Chapter 12

Understanding Time Travel

We start this chapter with the understanding that there can be no such thing as traveling back in time. However when we look at light coming to us from space, we know that light took time to reach us so in a sense, we are looking back in time. Considering the size of the Universe, the speed of light is considered relatively slow as it takes so much time to cross the vast distances. But is it really that slow? This chapter will show how time is actually much shorter for light travel across vast distances of the Universe. This solution is based upon the building blocks of physics itself. We have reviewed the building blocks in previous chapters defining the properties of gravity and energy as it exists in the Universe created by God.

Postulate U: The Principle of Duality of Energy and Light: Energy and light created within a nuclear reaction of a star will respond in two distinctive ways: (1) Massless energy will accelerate outward from the star at infinite velocity escaping the stars expanding gravity field and continue to radiate outward at infinite speeds until its potential becomes ineffective due to the inverse square law. (2)The duality of energy in the presence of gravity will also produce visible light which will travel along with gravity waves at the speed of light outward from the star.

Lets break this down a little so we are clear with the postulate. This dual state of existence is supported by the *2nd Postulate of Relativity* [32] which states *"The speed of light in a vacuum is the same for all observers, regardless of their motion relative to the source."* Therefore, light is always traveling at the speed of light even if the observer is traveling at the speed of light. We followed this step process earlier with the *angels* as they accelerated up to infinite speeds due to existing in a massless state. This state of energy will weaken as it extends out in proportion to the inverse square law [31] of energy and gravity.

We easily think of light as traveling at the speed of light but gravity actually does the same thing. Gravity produced in a star or any other body of mass "radiates outward" and also travels at the speed of light.

As we see stars today, they are generally in a steady state producing energy and light, so a flash or supernova or disturbance is needed to establish a time of occurrence. Then, applying the laws of energy and light in a *Created Universe*, we can determine its distance and time of occurrence. If we were to witness a supernova at say 1 KLY away, we would know it happened one-thousand years ago. This works because we are within the same gravity field or bubble. We will look at what happens when the supernova is greater than our current 6 KLY distance because gravity-free space changes things.

As stated in the postulate, if the supernova's gravity is greater than our 6 KLY gravity bubble, then its light will accelerate outward at infinite speed until it reaches our gravity field. It reacts with our gravity by slowing down and becoming visible. Then, the light coming toward us is actually traveling *opposite* the direction of our gravity field. It can only do this if the 2nd Postualte of Relativity holds true. To travel opposite gravity waves, it must travel at twice the speed of gravity. Lets look at some drawings below to explain this better.

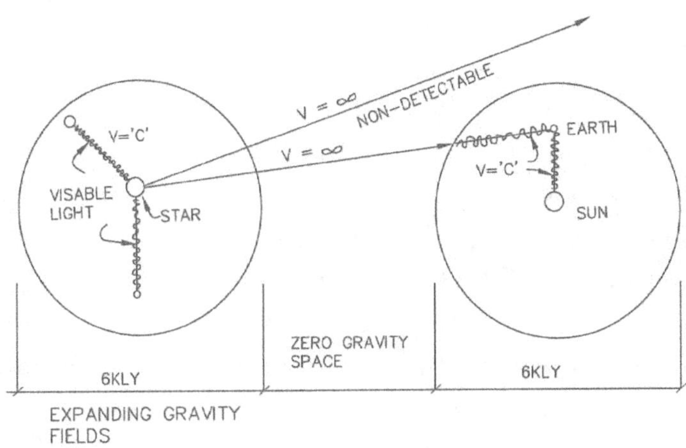

Figure 29: The Principle of Duality of Energy and Light

Figure 29 shows that gravity reacts locally at speed 'c' with energy in both gravity fields. Also, the star's energy will accelerate to infinite speed until it reaches our sun's gravity where it slows down and is observable. Energy traveling at infinite speed cannot be detected.

The Universe is an Energy Highway Network

We know from physics that energy is massless and should not react with gravity but it does when in its presence. From the postulates offered in this work, we proposed that energy traverses the Universe at speeds unrestricted by the speed of gravity. Since there were no expanded gravity fields at the time of Creation, energy from nuclear reactions of every star instantly sped ahead of gravity throughout the Universe, so light of every star would have been observable everywhere at the same time. Six-Thousand years later, energy continues to escape gravity and streams from each nuclear reaction. Each year, it takes distant star light a year longer to reach us due to our gravity field's rate of expansion.

So we claim the Universe is a highway network as it has always been filled with energy that has existed everywhere at the same time. However, only energy arriving at the boundary of an expanding gravity field travels at the speed of light. We learned earlier that " *As light travels from source to detection, it can behave as either a particle or a wave—or, paradoxically, both states or neither state."* So our claim and explanations do not disagree with physics as light can exist in two places at the same time because of its dual relationship.

Determining the Time of Occurrence of a Supernova

Contemporary Cosmology Calculation:

This calculation is simple if we assume the distance to an event is known. Since distances are light-years (LY) and speed is the speed of light "C", any distance light travels "in our gravity field" takes that many years, so to travel six-thousand light-years distance, takes six-thousand years.

But what if we claim a supernova is 13 billion LY (13 BLY) away? Contemporary cosmology says it takes 13 billion years for the light to reach us. This is assuming the Universe is that old. Of course Creation Cosmology denies that claim.

Creation Cosmology Calculation:

The following calculations have nothing to do with the physical distance between the objects but everything to do with the distance of our expanding gravity field. Since Creation was about six-thousand years ago, the expanded gravity fields of our sun and every star is equal to 6 KLY. Two-thousand years ago, the expanding fields would have a radius of 4 KLY, two-thousand years from now, the expanding fields would have a radius of 8 KLY.

The following Table is based upon the expanded size of the gravity fields at the time of a supernova. If a supernova distance from our sun is 20 KLY then how can the Universe be only six-thousand years old? It is so because the distance is based solely upon the radius of the expanding gravity field of our sun. Since our gravity field is constantly expanding at the speed of light "c", we must take into account the radius of the gravity field at the event time.

Lets see if we can resolve this situation. We start by reminding ourselves of the *principle of duality* of energy and light. Since that is the case, the energy radiation produced by the supernova expands throughout the Universe instantaneously. The phenomena of gravity acting upon energy radiation within a gravity field makes the energy visible to an observer. The energy that expanded at infinite speeds contacts our gravity field and becomes visible at the limit of our expanding gravity field but must travel in wave form generated by gravity to the observer and that takes time.

The light is also traveling against our expanding gravity field so it must overcome that speed C' and then travel at the another speed C' in order to reach us (so that's (2C). We can only observe light speed at 1C.

Time from (t=0) plus time of light travel through gravity
(t = r/c)   t = years,   r = radius (LY) of expanding shell,   c = speed of light in gravity

Estimated Distance to Supernova	Time (t) of Supernova from (t = 0) Years after Creation	Radius of our Sun's Expanding Gravity Field in Light years (r) at Time of Event	Light Speed = 'c'	Cumulative time from (t=0) plus c' time in gravity	Date Event Seen on Earth (Gregorian Calendar)
Distant Supernova					(approx)
13 Billion LY	0 (Creation)	0	c'	0	4000 BC
13 Billion LY	1 yr	1 LY	c'	2 years	3998 BC
13 Billion LY	10 yrs	10 LY	c'	20 years	3980 BC
13 Billion LY	100 yrs	100 LY	c'	200 years	3800 BC
13 Billion LY	1000 yrs	1000 LY	c'	2000 years	2000 BC
13 Billion LY	2000 yrs	2000 LY	c'	4000 years	20 AD
13 Billion LY*	**3000 yrs**	**3000 LY**	**c'**	**6000 years**	**2020 AD**
13 Billion LY	4000 yrs	4000 LY	c'	8000 years	4020 AD
13 Billion LY	5000 yrs	5000 LY	c'	10000 years	6020 AD
13 Billion LY	**6000 yrs (today)**	**6000 LY**	**c'**	**12000 years**	**8020 AD**

* light traveled instantly for a distance of all but 3000 LY

SN 1054 Crab Nebula	Time (t) of Supernova from (t = 0) Years after Creation				Recorded in 1054 AD
6300 LY**	2527 yrs	2527 LY	c'	5054 years	1054 AD

Figure 30: Table showing date light of a supernova and the Crab Nebula would be seen on earth.

Figure 30 shows the time and date a supernova would be seen and the time a future event would be seen according to our calendar. The Crab Nebula event would have occurred 2527 years after creation since it was recorded in 1054 AD. That was 966 years ago (6000-966) = 5054 years which is 1054 AD calendar date.

The Expansion of Gravity and Light Interference

What else happens to light we observe in the Universe? Since our gravity field has expanded out to six-thousand light-years, distant light must also travel though gravity fields of other stars. All observations of light coming to us from deep space is the result of the summaries of all of the bubbles adding red-shifts as the light passes through on the way to us.

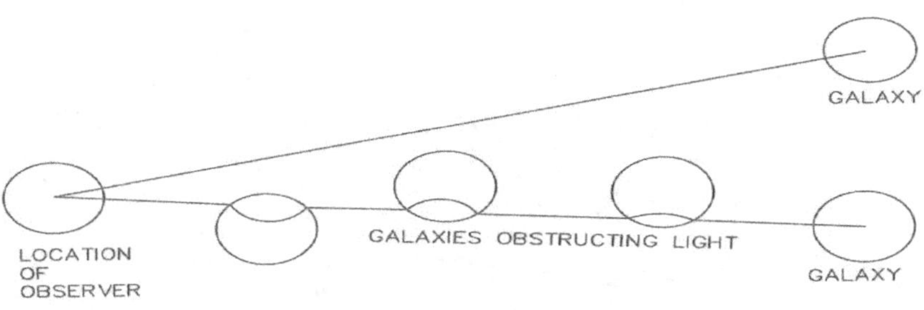

Figure 19c: Interference of light adding red-shifts coming to Earth

When we observe light coming to us, we do not know whether the light traveled through other gravity fields or if it came directly to us. The figure above indicates that other gravity fields can effect the light we see from stars and galaxies. *Figure 19c* shows why red-shifts alone cannot be relied upon to determine distances. Each time a light source passes through a gravity field the light is bent and that bending adds to the red shift distance. If distant light passed through several gravity fields, the galaxy would appear much more distant than a galaxy without interference even if they were the same physical distance.

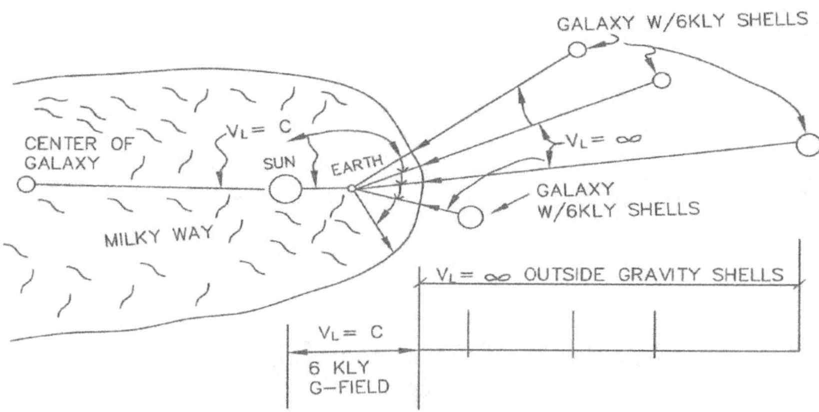

Figure 32: Represents today's gravity saturation within the Milky Way galaxy

Figure 32 represents today's situation with expanding gravity fields. There are thousands of stars currently within our 6 KLY bubble. It is now possible that our own Milky Way could be saturated with gravity fields so light within our galaxy will be observed acting only on the speed of light and gravity.

Testing the Micro-Gravity Speed of Light Theory

Postulate V: The interaction between energy and gravity will cease when the ratio of forces between them falls below a point of effectiveness. This ratio is stated as gravity to energy (g/e). Its values have not yet been determined.

The following is an "theoretical" example of testing for the speed of light under weakening gravity conditions. The test could determine if the speed of light accelerates as gravity weakens and what velocities would be possible as gravity fields approach zero. How do we actually test the postulate? First, consider launching a very small and very fast space craft beyond our solar system into deep space.

Let's look at how long it takes light to travel some distances at the speed of light (C) using Astronomical Unit distances (1 AU = distance from the sun to earth).

Sun to Earth = 1 AU. Light takes 8.33 min
Sun to 39.5 AU takes light 329 min (5.48 hrs)
          (*the edge of our solar system is 34.5 AU*)
Sun to 100 AU takes light 833 min (13.9 hrs)
Sun to 1000 AU takes light 8330 min (139 hrs)
Sun to 5000 AU takes light 41650 min (694 hrs)
Sun to 1 LY takes light 525,600 min (1 year)

If the craft could speed to ten percent of the speed of light, to reach each of the above charted distances, it would take ten times longer than light itself. To reach 1000 AU would take the craft about 1390 hours (57.9 days). That's still fast! If we could achieve this, we could test this relatively soon. (Time was rounded from 1388.33 hrs to 1390 hrs)

The smaller and lighter our test spacecraft, the sooner it will test the effects of gravity and give us answers we are seeking. Since this is a theoretical example, we will leave the detailed calculations for the engineers on our small spacecraft.

As our sun's gravity fields have expanded out six-thousand light years, we would not expect to escape our gravity bubble. The test is to determined if gravity could become so weak it would no longer affect the craft or its signals. That's what we are hoping to find out. As we saw in previous chapters, actual energy wavelengths may diminish and flatten as gravity forces weaken. The ratios of gravity to energy may no longer influence the speed of energy and light.

Unfortunately, at contemporary spacecraft velocities, it would take about 18,500 years to reach one light year. However the craft in our test is a very small dedicated only to reaching one light-year in just ten years. Practically speaking, we can not reach a speed where we can overtake our expanding gravity field so theoretically we will always be under influence but in very weak fields.

Who knows, we may someday be able to produce cancellation of gravity waves allowing us to further test this theory. Some claim aliens have already done this but for now, let reason prevail. Going forward with our example of reaching ten percent of the speed of light, lets see what such an experiment might yield provided our postulate is true.

If our experiment holds true, we could expect two possible outcomes: As the sun's gravity becomes so weak, energy and light speed accelerates in some proportion to the decrease in gravity density. Or: As gravity reaches some minimum threshold approaching zero, the speed of light accelerates immediately to infinity. The following diagram will help visualize these tests.

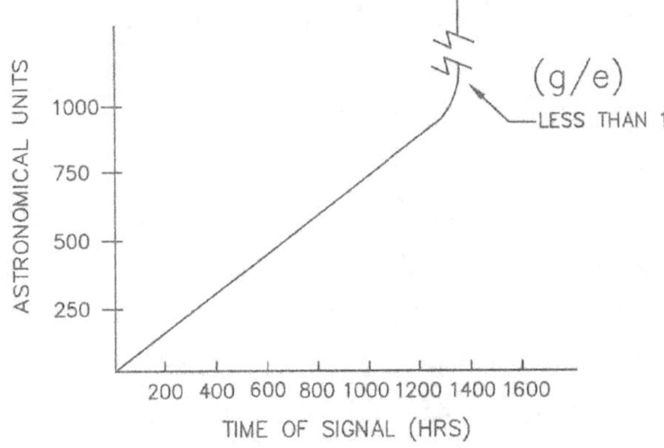

Figure 33: Time of signal if craft velocity is (0.1 C)

Once our craft reaches a constant velocity, it is designed to send back an identifier signal pulse at every astronomical unit. As the craft proceeds, each signal will take longer and longer to reach us. Each AU adds 83.3 minutes of time to reach us. As the craft journeys, we expect each flash we receive to be directly proportional to the distance as seen in Figure 33. Lets say that at the 1000 AU mark, our periodic signals began to change. Figure 33 shows that if (g/e) falls below some threshold, all future signals would also occur at 1390 hours.

An empirical reasonable and calculable conclusion of our sun's gravitational field tells us it is so weak out there it can no longer constrain the speed of light. If the craft signal actually escaped the gravity influence of our sun at 1000 AU , the signals will from then on, would occur exactly at the 1390 hour mark. At any greater distances, the signal travels at infinite speed back to the 1000 AU(g/e) limit. Of course the 1000 AU is only for the example. The change could occur at a much earlier or later time or it may never occur.

At each previous AU marker it emitted a high energy signal pulse back toward earth and we detect and record it. The spacecraft contains a highly accurate atomic clock that was carefully synchronized with an earth-based clock. We are going to ignore any claimed time differences due to Special Relativity (or gravity bending light). The 1000 AU flash was detected at 1390 hours as it responded perfectly to the speed of light in a gravitational field. At 1001 AU, the flash was also detected again exactly at 1390 hours. It did not add the extra 83.3 minutes. At 1002 AU, it too, occurred at 1390 hours and all future signals occurred at that time. This is a clear indication that our craft has escaped the effects of gravity acting on the speed of energy and light. There could still be some variations due to the gravity of the craft itself and that could affect the time.

Conclusion

As the craft continued, each consecutive flash would also occur at 1390 hours. This experiment would prove or disprove the postulate that energy and light will instantly traverse micro-gravitational space as there is absolutely nothing to resist, slow or stop it until it reaches a sufficient gravitational field where it again reacts with gravity waves and produces light.

Based upon known laws and physics about the different states of existence of light and energy. These explanations are logical and contrary to contemporary claims for the time required for light to travel across the Universe.

Chapter 13

Anomalies in Space

Colliding Galaxies

How do we explain two colliding galaxies or is it necessary? If two galaxies are created at the very same moment in the very same space, it is apparent that in time, they would react in some way. We know that star light emitted behind another start can be bend, disperse and distort its image like we see with *Einstein Cross*. We do not see stars in their actual location. An experiment using two images through two layers of curved glass can distort images significantly. This would create the illusion that two galaxies may be in collision. The point is that light can distort images skewing them to appear significantly different than they actually are.

This was touched on here as colliding galaxies may represent a valid argument for longer time. However, a lot could happen to images of two galaxies so close to each other in six-thousand years. The effects could include mutual gravity, relative motion and light distortion caused by refraction. We have only observed these over the past one-hundred fifty years.

Formation of Spiral Galaxies

There does not appear to be any scientific explanation for how galaxies came to spiral.Therefore, the following explanation starts with an assumption just as assumptions are made in contemporary cosmology. This assumption is that the Creator gave the galaxies an initial spin. Cosmology has not explained galaxy spin so this is just as valid. This brings to question the age of the Universe. In a big bang universe, how did the building blocks of basic matter gather into clusters of galaxies in the first place? The notion that most matter in the Universe decided to clump into spiral galaxies does not seem to follow any rule or order, especially in a void universe filled with randomly distributed matter. Logically, all of the matter in the Universe should have gathered into one enormous ring galaxy, or at least still be in the process. One could also claim since the Universe is now expanding faster than previous, the Universe could not clump into one mass. What caused galaxies to form, rotate and accelerate? We do not know.

Lets see what may have happened in an old universe. Through some yet unexplained physical reasons, stars formed and then they formed themselves into galaxies. Each of the millions of stars having their own gravity fields attracted themselves toward each other and mysteriously began to rotate. The stars are attracted to each other so some mutual gravity attraction and rotation tends to draw them together into bands. Eventually, the central gravity gradient builds creating stronger G-forces and begins to draw the outer bands. The outer bands appear to be traveling faster than they should when compared with the complete spiral and central parts of the galaxy. None of this makes sense when applying the rules and principles of physics. This is still a mystery.

A Simple Shop Experiment

The following is a simple shop experiment but the results tend to match the observed rotation of the spiral arms within galaxies. Figure 34 shows a drill motor with a magnetic rod in place of a drill bit. The plastic tray is glued to the bottom of the magnetic rod so it will rotate. Iron filings are randomly placed in the tray and are attracted to the magnet.

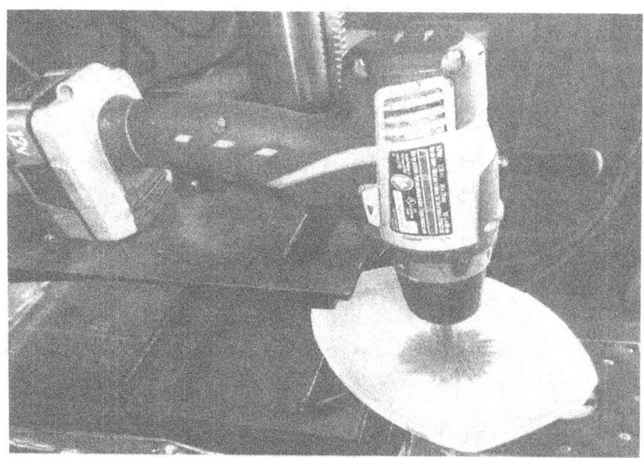

Figure 34: Spiraling Magnetic Fields

Figure 35 below shows the results when the drill motor is spun in the *clockwise direction* at about 600 rpm. The filings tend to group and form leading bands toward the direction of rotation. This seems the opposite of what we should expect.

Figure 35: Iron filings spun in clockwise direction

Figure 36 below shows the results when the drill motor is spun in the *counter-clockwise* direction. The same filings tend to regroup and form leading bands in the direction of rotation. Again, just opposite of what is expected.

Figure 36: Iron filings spun in counter-clockwise direction

While the test is simple and interesting, it did not tell us anything about how galaxies came to rotate but it does show that with rotation, the leading edges tend to point in the direction of rotation rather than lag and point back. The drill motor is an outside effect and not an internal effect. So galaxy rotation must have been caused by some universal outside effect.

The Case Against Cosmic Background Radiation [33]

We will discuss two possible forms of energy radiation being detected as CMB.

The young Universe is filled with energy in the form of radiation (noise) that is not in wave form and not detectable outside of gravity shells. Every star and every galaxy are producers of energy. We should be able to detect continuous streams coming to us from our own Milky Way as our gravity fields continue to expand.

This continuous expansion is also bringing together multiple gravity fields of stars within the Milky Way. When gravity fields combine, the radioactive particles can then transfer from one gravity field to another. So our sun is receiving a continuous flow of radiation from our nearest neighbors.

Energy that has escape gravity throughout the Universe is massless undetectable energy that may be defined as energy vectors radiating out in every direction at infinite speed. We should be continuously receiving energy being converted into detectable forms of radiation from everywhere in the Universe. The only restrictions are based upon the radiation strength as it dissipates in proportion to the inverse square law.

Cosmic background radiation (CMB) [33] is a form of energy. At this moment, our universe is alive and functioning as intended. Incalculable amounts of energy and light are being produced in our three-dimensional universe. If our universe is not old, the cosmic background radiation experiments must be detecting radioactive particles left over from nuclear reactions of nearby stars.

If there was a Big Bang, how would disproportional radiation (energy) still exist for billions of years and be spread randomly. To single out from all of the universal energy produced, such insignificant amounts of energy claiming it is from something that happened 13.6 billion years ago is very very thin to say the least. We need to answer these questions: How fast would that energy decay? Was the Big Bang a super-intense radioactive singularity spewing that much radiation into the universe? What type of expansion was the Big Bang in the first place? The reality is that we know nothing at all about any of this and cannot use cosmic background radiation to solve it. It remains a mystery.

Consider a flash of light passing you by. Once the energy of that light has gone by, a reflecting surface is required to send it back to you. It is not reasonable to consider that light or energy could continue to bounce and reflect throughout the universe over billions of years. Its energy would very rapidly dissipate. If there was cosmic background radiation in the universe from a big bang, it would be long gone. Cosmic radiation must be a recurring source from nearby stars within our own Milky Way.

This background radiation is the result of our Sun's expanding gravity field merging with other nearby expanding gravity fields and may be picking up the noise in the form of radioactive particles due to those mergers. If that is the case, then gravity-free space may be acting as some type of filter for low level energy so it can only be detected when gravity fields merge.

Another consideration is that if cosmic background radiation actually exists today, it would have been exponentially much stronger in the very recent past. This would have potentially killed all life on an earth if the earth was billions of years old. A lot of exculpatory information does not support the theory for cosmic background radiation!

A perfectly ordered Universe vs one like ours?

The Universe can be compared to a painting created by a famous artist. The painting is for viewing and was never intended to be something we could bring to life or interact with. However, the Creator is someone we can interact with. We can examine his motives and question his techniques. Even though the Creator has limited his interaction with us and an artist may have long since passed away, we can research their works and historical records to learn more about them. The painting itself is like the observable Universe, neither can be touched but they can be carefully examined.

The *Mona Lisa* shrouds us in mystery about the girl and the smile. Da Vinci did not reveal to us his purpose for that particular painting so we look upon it with a certain intrigue and mystery. A thousand stories could be made up about her. The Creator's Universe is no different. We were never intended to travel through it as it is physically impractical and impossible. Does this exclude us from studying it? Like the *Mona Lisa,* we should be inspired by its grandeur and wonder. Who among us watched Da Vinci create the *Mona Lisa*? Who among us watched the creation of the Universe?

Actually there still is one among us who left us a pretty good account of his work that stands much better than anything the mind of man has come up with. Scriptures define Christ Jesus as our Designer and Creator. He is alive and he is returning to collect those who have accepted his promises by faith. Just as we should love, honor and respect our parents and our children, we should love our Creator and understand that it is by his will we are given life. If we choose to ignore our Creator, he will ignore us. Much greater than this life is his promise to those who believe. It is an eternal life in much better bodies. If you truly want to find out the answers to all of the mysteries of life, then I encourage you to attend the meetings to be held in Heaven.

If this Bubble Theory is correct, how did we get so misled?

There are many great resources available for understanding science from a Creation perspective. If we take the time to unravel Darwins evolution using contemporary knowledge of science, we find life is much too complex and complicated to have ever evolved. Yes, there is variation among species and there is much information designed and built into our DNA which has allowed us some flexibility to adapt to environmental conditions. This adaption is much more prevalent in animals and plants than it is in mankind. This was all by design.

Our Creator did not include that level of variation into the DNA of mankind whom he chose to fellowship with. Over a few thousand years, mankind has maintained his basic biology. This alone says something about the design of nature and not about evolution of nature. From the time of Adam and Eve and then from the time of Noah and his family, we are still one people with minor adaptions to our environment. Unfortunately it has been our cultures and ideologies that divide us. Race has nothing to do with it as we are all one people. God loves each and every one of us and Heaven rejoices every time one of us accepts God's love in return.

Conclusion

Looking at galaxies, we know that galaxy formation, rotational velocities, centripetal forces and appearances do not make sense without some external organizing forces. Radical theories have been developed looking for some other force or material in space to compensate for this abnormal rotational velocity phenomena where the outer rings rotate faster than expected. An explanation is plausible when considering an outside influence or a Creator.

Chapter 14

Bringing God into the Equations

So How Did God Create the Universe?

What I am about to present is of course speculation just as contemporary cosmologist speculate their view of how the Universe came to be. This speculation is based upon several facts about physics, Holy Scriptures and faith as presented throughout this book.

We learned from physics that matter can be converted into energy such as in a nuclear reaction. We also learned that when light enters a gravity field or is created within a gravity field, photons are produced by gravity and they actually perform physical work. So we see examples of matter-to-energy and energy-to-matter conversion in both directions. First matter into energy in a nuclear reaction and energy into photons and other usable forms such as heat in the other direction.

Just as our DNA contains all of the information needed for our human development from conception to adulthood, our Creator God has also placed intelligent design into his energy. His creation involved conversion of energy to matter and that consequently produced gravity.

Our Creator God initially created energy from resources belonging to him and laid it out in order throughout a vast void he had set aside for his Creation. God performed his Creation work over a six-day earth-time period. Starting with step one, the massless energy was gathered into intelligent modules he designed and created and were spread out into the void. On *day-one*, the Universe was filled with carefully laid out modules of massless energy.

This intelligent energy was still in an undetectable state just as energy traversing zero gravity space. In an instant, God introduced an action that transformed all of the carefully placed modules of energy into bodies of mass. The Universe came into physical existence. Then God said "let there be light" and all of the bodies in the Universe appeared through this energy-to-mass conversion and the nuclear fires lit.

The processes continued each day on earth by making all things he designed using forms of energy and converting them into bodies of mass. Then with Adam and Eve, he breathed his spirit energy into them making them living souls.

There is absolutely no way to over-state the power of our Creator God. He is capable of doing such things. Remember from earlier that like the movie producer, all things were carefully thought out in advance before his Creation began.

What God has done is known to physics because of our knowledge of energy-to-mass conversions and mass-to-energy conversions. However, he did it on a much grander scale. *Figure 37* shows how the power and resources of our Creator God can be explained in a way for us to imagine his genius.

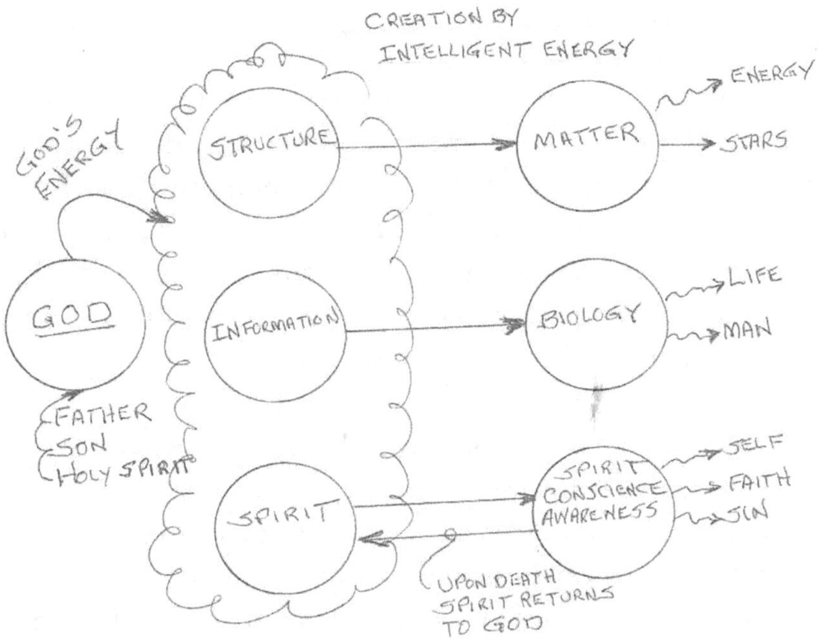

Figure 37: Theory of God's use of energy to create

Following His death and Crucifixion
How was Jesus able to appear to his disciples and others?

God used energy to create the Universe and all that is in it. We will now look at such things on a smaller scale but just as great. We begin by coupling the intelligent Spirit of God with intelligence God built into energy allowing it to transform into so many different things. To give you an idea of how this coupling can be, consider the contents of this book transmitted wirelessly from earth to the moon then printed out on paper. The process went from physical existence to energy in the form of electro-magnetic radiation then back into physical existence but on the moon. That energy contained information.

The ability for Jesus and some angels to come to earth from their relm in Heaven simply requires their massless spirits in the form of energy, to be placed at a precise location, then the energy-to-mass conversion power of God takes place and here they are. This explains how Jesus, Gabriel, Michael and other angel messengers from God were and still are able to appear on earth then return to Heaven.

Scriptures tell us we will have physical bodies during our time in Heaven so Jesus must also. Just as printers and transmitters can convert objects back and forth, so can God with our bodies. Regardless of our physical states, God has us on record. He knows us as we are spirits housed in physical bodies here on earth. Of course there are other angels not from God who can and have appeared on earth.

Sometimes, understanding the power of God requires us to apply what we know from science in logical ways, especially for the scientific or inquiring mind.

Blessed are those who believe simply by faith; Jesus told Thomas: *"Then Jesus told him, "Because you have seen me, you have believed; blessed are those who have not seen and yet have believed." John 20:29.* I have believed by God's calling and by the empirical wisdom God has given me, yet I have pursued greater understanding and found profound truth in God's Holy Word.

John 20: The Empty Tomb

1Early on the first day of the week, while it was still dark, Mary Magdalene went to the tomb and saw that the stone had been removed from the entrance. 2So she came running to Simon Peter and the other disciple, the one Jesus loved, and said, "They have taken the Lord out of the tomb, and we don't know where they have put him!"

3So Peter and the other disciple started for the tomb. 4Both were running, but the other disciple outran Peter and reached the tomb first. 5He bent over and looked in at the strips of linen lying there but did not go in. 6Then Simon Peter came along behind him and went straight into the tomb. He saw the strips of linen lying there, 7as well as the cloth that had been wrapped around Jesus' head. The cloth was still lying in its place, separate from the linen. 8Finally the other disciple, who had reached the tomb first, also went inside. He saw and believed. 9(They still did not understand from Scripture that Jesus had to rise from the dead.) 10Then the disciples went back to where they were staying.

Jesus Appears to Mary Magdalene

11Now Mary stood outside the tomb crying. As she wept, she bent over to look into the tomb 12and saw two angels in white, seated where Jesus' body had been, one at the head and the other at the foot.

13They asked her, "Woman, why are you crying?"

"They have taken my Lord away," she said, "and I don't know where they have put him." 14At this, she turned around and saw Jesus standing there, but she did not realize that it was Jesus.

15He asked her, "Woman, why are you crying? Who is it you are looking for?"

Thinking he was the gardener, she said, "Sir, if you have carried him away, tell me where you have put him, and I will get him."

16Jesus said to her, "Mary."

She turned toward him and cried out in Aramaic, "Rabboni!" (which means "Teacher").

17Jesus said, "Do not hold on to me, for I have not yet ascended to the Father. Go instead to my brothers and tell them, 'I am ascending to my Father and your Father, to my God and your God.' "

18Mary Magdalene went to the disciples with the news: "I have seen the Lord!" And she told them that he had said these things to her.

Jesus Appears to His Disciples

19On the evening of that first day of the week, when the disciples were together, with the doors locked for fear of the Jewish leaders, Jesus came and stood among them and said, "Peace be with you!" 20After he said this, he showed them his hands and side. The disciples were overjoyed when they saw the Lord.

21Again Jesus said, "Peace be with you! As the Father has sent me, I am sending you." 22And with that he breathed on them and said, "Receive the Holy Spirit. 23If you forgive anyone's sins, their sins are forgiven; if you do not forgive them, they are not forgiven."

Jesus Appears to Thomas

24Now Thomas (also known as Didymus a), one of the Twelve, was not with the disciples when Jesus came. 25So the other disciples told him, "We have seen the Lord!"

But he said to them, "Unless I see the nail marks in his hands and put my finger where the nails were, and put my hand into his side, I will not believe."

26A week later his disciples were in the house again, and Thomas was with them. Though the doors were locked, Jesus came and stood among them and said, "Peace be with you!" 27Then he said to Thomas, "Put your finger here; see my hands. Reach out your hand and put it into my side. Stop doubting and believe."

> **28**Thomas said to him, *"My Lord and my God!"*
>
> **29**Then Jesus told him, *"Because you have seen me, you have believed; blessed are those who have not seen and yet have believed."*

The Purpose of John's Gospel

> **30***Jesus performed many other signs in the presence of his disciples, which are not recorded in this book.* **31***But these are written that you may believe* b *that Jesus is the Messiah, the Son of God, and that by believing you may have life in his name.*

New International Version (NIV)

Holy Bible, New International Version®, NIV®
Copyright © 1973, 1978, 1984, 2011 by Biblica, Inc.®
Used by permission. All rights reserved worldwide.
Bible Hub

I have included this description of Jesus' appearances after his crucifixion. For those who do not know the full story of the life and death of Jesus, biblical history and testimony leave no room for picking this apart. Jesus was crucified and died and was placed in a sealed tomb. He was risen, telling Mary not to hold on to him as his physical state was different than his earthly body. Later, he returned after ascending to the Father. This time his body was restored into a physical body with the wounds from the spikes and the spear through his side. The point is that God has the knowledge and power to Create and to resurrect his believers taking them with him back to Heaven. It becomes a matter of what we are willing to do with this knowledge.

Movie screen analogy:

When we look into space on a clear dark night, we see the wonders of the heavens. It is like looking at a giant 3-D movie screen placed all around us. Energy from the distant reaches of space contacts that screen so the energy coming from our universe is displayed on the screen as visible light.

Let's say we have been watching this screen for a very long time, in fact, since Creation, so we see first light striking our screen. The screen seems to have come alive as star lights all appeared on this dark night. Because we know something about space and time, we come to realize that the energy for this light traversed great distances very quickly. This has been an age-old mystery fully credited to God up until about three hundred years ago when mankind introduced the idea that our Creator was not involved.

This screen analogy is interesting in the same way, according to the Bible, there will be a time when that screen will be opened and our Creator will be revealed. Jesus Christ who is God in human form is coming back to remove Believers and to punish those who do not believe.

Let's examine how we would approach verification of the postulates presented here.

If you are well informed on the subjects of cosmology and physics, you will know there are many areas of the science that were not covered in this work. We covered several relevant fields. We needed a place to start in proving this theory. We cannot remove ourselves or our laboratory from the presence of gravity, but we are resourceful people and can develop ways to find out new things about Cosmology and

Creation. I say "open the doors to new understandings". This involves a new starting date for the "creation" of the Universe and the earth.

If the Holy Bible is true, and I believe it is, God, our Creator has told us that when mankind turns away from Him, his fate is sealed in the second death. Believers are appointed to die only once. A Believer is one who has accepted God in faith as our Creator. This work is to show how easily we can be deceived when we take contemporary science's view of our beginnings over God's truth.

Isaiah 34:4
All the stars of heaven will be dissolved. The skies will be rolled up like a scroll, and all their stars will fall like withered leaves on the vine, and foliage on the fig tree.

Revelation 6:14
New International Version
The heavens receded like a scroll being rolled up, and every mountain and island was removed from its place.

Conclusion

My own hand laid the foundations of the earth, and my right hand spread out the heavens; when I summon them, they all stand up together. Isaiah 48:13

God created!

Chapter 15

Facts on Trial

There are three main areas introduced in this writing that collectively cause us to focus on the real possibility that pure energy reacts differently in gravity than it does in non-gravitational space. Einstein's Special Relativity assumes the existence of gravity, but in a young universe, the vastness of space contains no gravity and that opens a whole new set of rules. We learned that Relativity is actually based upon gravity fields bending light causing it to travel longer distances and that causes light to shift toward the red spectrum and that adds more time for light to travel through gravity.

We learned that the present day state of cosmology, is filled with innuendo and speculation which in itself is reason and justification to look in other places to find answers about our beginnings. We also learned that the credibility of the Holy Scriptures cannot be ignored as a starting place to finally solve not only our existence but the reasons why we exist.

Why Does Proving a Young Universe Matter So Much?

Let's respond to some commonly read or heard secular statements and questions. We will respond from a Biblical Creation point of view:

"There is no god because we evolved." Anyone who makes this claim is no different than having faith in *naturalism* and in an old universe of unproven beginnings. This claim rejects Judeo/Christian Scriptures and many other religions believing in a Creator.

"Aliens from other worlds visit the earth and put us here." Faith in aliens kicks the can of reason down the road when looking for reasons for our existence. Aliens other than heavenly visitors are not referenced in the Scriptures.

"Ghosts of the dead exist and appear among us." Ghosts walking among us denies Scriptures telling us that when we die, we all rest until the return of Christ who will judge us. However, there are spiritual and angelic beings who move among us, not ghosts.

"Humanity is basically good." Believing humanity is basically good is contrary to Scriptures telling us that mankind is fallen from the grace and love of our Creator. This requires action on each and every one of us to establish individual relationships with the Savior in order to be rescued from this world.

"Miracles are impossible." Miracles are well documented throughout history, some supernatural actions are required for miracles so supernatural forces must exist.

"Angels do not exist." Angels are well identified in Scriptures and are both good and evil; those who followed Satan are separated from the love of God just as Satan. Angels are created beings. Those from God serve us as messengers. Those from Satan serve as destroyers and deceivers.

"Spiritual forces have nothing to do with us." Spiritual forces are inherent and influential in our minds causing us to do sinful and evil things especially when we are spiritually weak and ignorant of the love of God.

"This book came from a singularity." Believing that this book came from a singularity or a big bang requires faith beyond all reason. To believe that, requires us to believe that information came from no-information or something came from nothing.

"Mankind learns and benefits from history." Mankind has learned to repeat history by committing the same evil acts upon humanity over and over again. We learn nothing when we give our lives over to the influence of sinful men. Sin is already deeply embedded into the nature of mankind, so external spiritual influences are no longer required.

"Human life has no meaning." One who believes that human life has no meaning is a hypocrite because he holds to the protection of the good that is built into Judeo/Christian laws founded in most democratic nations. If life has no meaning, then mankind's sinful and selfish nature becomes driving forces behind all forms of government and that leads to tyranny and that suppresses the God-given right to freedom of human life.

"There is hope that mankind will evolve to a higher state." Believing that we are still evolving to higher states does absolutely nothing for us as we are here in the now. According to the Bible, there is no long-term future for mankind due to his rebellion away from his Creator. Rescue from this world comes only through faith in God as our Creator. Holy Scriptures tell us there is a price to pay for the fallen nature of mankind. Believers are only forgiven because a Savior paid the ransom demanded by God, for our sins. Accepting the price paid by Jesus Christ comes with a caveat, it requires something very simple. We must come to the realization that God created us and loves his Creation. If we accept that by faith, then we become children of God and with that comes a new eternal life. The Bible is the most profound book in the world as it explains all of this for those of us willing to trust and accept his Word by faith. In the beginning was the Word known as Jesus Christ.

"There is no inherent knowledge of a Creator." There is inherent knowledge of the supernatural as evidenced by all of the above reasons; Scriptures tell us exactly who this supernatural being is. The Bible is logical, historically factual and accurate and it foretells the future of mankind up to the ultimate judgment and destruction of this world.

"I'm happy with those making the rules I live by." There is an unrecognized blessing upon those reading this book since the rules most of us are now living by were designed by those who had faith in God our Creator. Those rules were for the mutual benefit of mankind to be free to have such a relationship with our Creator. We should want to preserve those freedoms and relationship and not change it by the will of sinful people.

"I'm okay with atheist making the rules." Should anyone be okay with atheist making the rules we live by? Those who declare themselves as atheist seem to have a vendetta against their Creator which brings to question the influence of their motivations and what forces are behind them. Evolution is the product of atheism and agnosticism.

"Public indecency is okay for my children to see." The Biblical accounts of Sodom and Gomorrah tells us of a time when public indecency was rampant with mankind given over completely to his sinful nature. The only thing standing in the way is good men willing to do something as oppose to "good men doing nothing". The spirit of good men standing in the way is the Spirit of God within us telling us those things are wrong.

All of this brings us to the question of "what really matters in our lives and what is our destiny?" The Holy Bible provides us with a very good answer. *"Just as people are destined to die once, and after that to face judgment. "so Christ was sacrificed once to take away the sins of many; and he will appear a second time, not to bear sin, but to bring salvation to those who are waiting for him." Hebrews 9:27-28*

Understanding this verse is to know Christ the Savior and what he has done for those willing to accept his gift of eternal life. Scriptures tell us there will be a judgment and second death for those who reject our Creator God

The final question: *"Is the Universe actually about six-thousand years old?"* Answer: Yep!

Conclusion

33) When they came to the place called the Skull, they crucified him there, along with the criminals—one on his right, the other on his left. 34) Jesus said, "Father, forgive them, for they do not know what they are doing " Luke 23:33-34 Holy Bible NIV (BibleHub.com)

This project is also dedicated to the millions, perhaps billions of people, throughout history whose lives and freedoms have been taken from them by those who *"do not know what they are doing"*. Mankind has fallen victim to countless numbers of godless tyrants, who have replaced the love and fellowship of our Creator God with their own selfish and evil ways. These historic destroyers have replaced God's Law with their own kind of law for the exclusive benefit of themselves. They desire sovereign power and control setting their own laws over mankind. In order to gain such power, they must remove all influence of our Creator God.

In so many subtle ways, the godless mindset of selfish men have lead to pain and suffering not only upon people of faith but upon lives of millions of innocents, who never knew the love of our Creator. God left all of us testimony and evidence of his Creation. When science teaches people to accept theories such as naturalism and evolution without taking into account the evidence of God's work, they mislead us and separate us from his relationship and purpose of our very existence.

I find it tragic that modern scientist *"do not know what they are doing"* because they are interfering with eternal relationships available to all who come to know and believe in our Creator. Those who interfere will face their own judgment by the one they have rejected. Our Creator cannot be ignored without consequences. But still, there is forgiveness in God's heart for all who come to him. This is a wake up call.

In case I have failed to convince you!

As a follow-on, our life expectancy is but a few years by comparison. The gospel of the "Good News" has been preached to all capable of reading this work throughout the world. Regardless of one's thinking of cosmology, it is extremely important to understand that we are a creation and therefore we have a Creator. Unless we understand the absolute need for a Savior from this sinful world and seek Him, we are doomed to a second death. If you do not know Him, let me introduce you to the Messiah, His name is many names but we know Him as Christ the Lord, He is Jesus Christ, our Savior.

No one doubts we are all appointed to die once. [25] You should not accept teachings of natural science but should open your eyes and expand the teachings in *Creation Cosmology 101-202*. *Creation Cosmology* is the establishment of faith in the God who has done and promised all of the things we find in the Holy Scriptures. *Naturalism* as opposed to *Cosmology 101-202* is a godless fabrication of our beginnings. Creation is based upon reason, logic, fact and faith. Lets seek the truth of the Word of God who told us what he has done for all to see. If we exclude the Word of God, we will never find the truth.

The following is a High School Report written by my 14 year old granddaughter, at the time. Relationships with the Creator depends so much on what our schools teach and what we commonly hear about naturalism and the age of the earth. Teach your children to be skeptical and to decide for themselves whether they arose from soup in a puddle a few million years ago.

Brooke Pugh

Mrs. Montinieri
English II Honors, period 2
May 18, 2009

Intelligent Design

Intelligent design is a controversial idea that contradicts the theory of evolution, that is supported by many documents and research that is believed to be the more realistic explanation of life on Earth.

Consistent with some scientists, to the "big bang" theory of evolution, the beginning of the universe was created by an intelligent agent using intelligent design. Pure modern evolutionary thought, based upon current scientific discoveries in the complexity of life make classic evolutionary theory extremely irrational and improbable. Supported by the IDEA Center, "Naturalist explanations are those which assume that there were no forces involved in history of life" (Intelligent). This explanation excludes matter, energy, and the physical and chemical laws of creation and sustaining life. It assumes there were no creators or intelligent influences, and does not explain the details of life, such as allowing life to flourish and exist. In Darwin's theory, it is said that "There appear to be few plausible examples of transitional forms in the fossil record, a major line of evidence. This counts against the naturalistic evolutionary transitions in the history of life"

(Intelligent). This infers that evolutionary theory does not go into detail whatsoever, while the creationism theory goes into minute detail as far as cells and their functioning antics. Without going into detail, this idea is irrational and incomplete. Additionally, scientists state, "As our understanding of the cell expands, it is becoming increasingly apparent that Darwinism is incompatible of explaining the origin of much biological complexity" (Intelligent). The significance of this quote explains that the biological structure of the human body alone is much too complicated to be implicated by evolving from single celled organisms. The irreducibly complex system of independently working organisms cannot be created by chance, as argued within the controversial topic of intelligent design versus evolution. In conclusion, according to the two different theories and ideas based on creationism and Darwinism, the history of the world is very different, one being inconclusive and the other being very detailed.

Intelligent design is known to be what Christians and all religions believe; and all atheists believe in evolution, when it really can work both ways. As avowed, "It's very possible that one could embrace the science of the Darwinian tradition and also embrace a Christian understanding of God at work in the natural world" (Christianity). The relevance of this is that to believe in evolution doesn't mean that you have to be an atheist. This is important because scientific research may reflect on the different branches of religion and the beliefs of different religions. As affirmed, "the Neo-Darwinian model of random genetic variation combined with natural selection provides the most adequate explanation for the development of life forms" (Christianity). This proves that people have different opinions of the evolution-creationism issue. It voices that the most reasonable explanation of life on earth is a combination of the two theories. Expressed here, "God has used the evolution of life over deep time to serve a divine purpose for creation" (Christianity). This means that religious and non-religious people can see eye-to-eye which changes scientific research and opinions. It formulates that the most rational explanation is also a combination of theories, God allowing for both. Theories may be changing when people start putting the two ideas together and calling it theistic evolution.

Thus, the theory of evolution and the idea of an intelligent designer is being put down and is almost wiped out. As spoken, "the displacement of God by Darwinian forces in this century is now almost complete" (evolution). This is not explaining the loss of God in society, but how an almost revolution of the way we look at things has taken place. Instead of fitting everything into God's will, we now look for the rational, natural forces on earth. Articulated here, "religion in general and God in particular, need not to be subordinated to the findings of evolutionary biology" (evolution). The importance of this quote is that it totally conflicts the thought of the displacement of God. It is saying that religion and God do not need to be subordinated because religion and evolution are not incompatible. An issue is proposed, "as a result, even if we were generous enough to accept science and religion as co-equal ways of knowing, Orwellian common sense would tell us that one of these partners would be more equal than the other" (evolution). The evolutionist's issue here is that science and religion cannot coexist. It is saying that one, science, would be much more dominant.. With the idea of evolution becoming stronger and more dominant, the idea of religion and intelligent design is becoming subordinate.

    The evidence of an intelligent designer is much stronger than that of evolution. In Darwinism it is believed that, "the secularization of science manifests itself in the belief that nature has no need for an intelligent designer but is self-caused and self-contained" (creationism). In the quote it is stated that there is no need for a designer, but everything somehow got here by itself. What this doesn't explain, is how everything got here and how these organisms became self-causing and self-contained. The contradictory complexity of life is explained as stated, "the very complex machine-like entities exist, which must be exactly as they are or they cease to function properly" (intelligent). This explains that life is so complex that if it is changed over time as believed, it would stop working as it should and die. The chances are too high for something to change so much and not cease to exist from a mistake of change. This quote contradicts Darwinism, "irreducibly complex structures cannot be built up through a Darwinian evolutionary process, because it says that a biological structure must be functional along every small step of its evolution" (intelligent).

This contradicts evolution because in "reverse engineering" of these organisms show extinction if they are changed even slightly.   This idea completely counter-opposes Darwinism because everything evolution is known for, changing over time, is completely impossible in the eyes of a creationist. In recapitulation, the laws of Darwinism and evolution absolutely conflict the beliefs of creationism.

There are two main explanations of life of earth: Evolution and Intelligent Design. Both are supported, but with more and more research, intelligent design is becoming the most rational looking answer to our ongoing question. In conclusion, these theories are very contradictory and both assume different positions on the issue of where life on earth came from and how it got here. Many either choose one side of the issue and shuns the other side. Apparently, many are now putting them together and making a new theory. With all of this information presented, it is still a personal decision that only an individual can make for themselves. Just know your theories.

<u>Summary of Postulates</u>

Postulate A: Energy and light can only be constrained by the forces and speed of gravity when in its presence.

Postulate B: God would have known the influence of sin upon mankind who would rebel and seek to circumvent His authority. God would have left evidence for the enquiring mind to find and justify his work. So we seek in earnest. The very survival of the Holy Scriptures over the millenia is a miracle itself.

Postulate C: Energy and light escaping gravity fields cannot conform to Newtonian Physics.

Postulate D: Contemporary scientific theories lack sufficient evidence to prove claims made by naturalism, evolution or a big bang. This lack of evidence postulates the door is wide open to examine and test other possibilities regarding our beginnings and the existence of our Universe.

Postulate E: Energy, including light is massless and must react with gravity in order to be seen or measured. Stated another way, massless energy such as light would not be seen or measured in the absence of gravity.

Postulate F: Units of light called "photons" may be defined as the product of energy in the presence of gravity. It is an unexplained occurrence or phenomena when combining massless energy with gravity. This combination produces harmonics or vibrations we understand as light waves. This phenomena converts non-material massless energy into useful work (or photons) of which life is dependent upon. A photon may only exist within a gravity field of sufficient strength.

Postulate G: Energy and light from the moment of production within a nuclear reaction accelerates radially outward at infinite velocity. However, energy and light can only be observed and measured at the speed of gravity within a gravity field.

Postulate H: The initial velocity of massless energy radiation is equal to infinity. The velocity of "usable" energy instantly changes to the speed of gravity in the presence of any new gravity field.

Postulate I: Since the speed of light in gravity shells has the same value for all observers regardless of their state of motion, the shortest possible time for light to travel is in a straight line. If light follows a longer path such as curving due to gravity refraction, we are simply adding the time necessary to travel a longer distance. Time dilation is nothing more than light traveling longer distances taking longer times to arrive.

Postulate J: The time required for light of an event such as a supernova to be seen can be derived by ($s = d/t$) or transposed to ($t = d/s$). The distance is the sum of distances where light travels through gravity fields: ($d_g = d_1 + d_2 + d_n ...$). Since 's' is the speed of light 'c', the equations becomes ($t = d_g/c$).

Postulate K: Light and energy from distant galaxies is refracted and dispersed in accordance with Newtonian physics. The laws of physics in a gravity-filled universe would dissipate energy so much, that distant objects could not be seen. In a young Universe mainly void of gravity interference, light will not refract or disperse as it would in a universe filled with gravity gradients. Light must necessarily travel through gravity-free space conserving energy in order to be seen.

Postulate L: Light from a background star or galaxy will refract producing an observable arc called lensing. Refracting light passing through the expanding gravity shell (bubble) of a foreground star will over time shift position or disappear due to the expanding shell as seen by an observer.

Postulate M: Gravity must possess an interactive relationship between itself and energy. Massless energy excited by gravity fields produces visible light and that responds to the design of the eye allowing it to see the energy as light. Without gravity, energy cannot be excited.

Postulate N: Strong gravitational forces must be present to produce Nuclear Reactions inside a star. Within a star's nuclear fusion reaction, each atom's reaction is independent of every other reaction occurring at billions of inharmonious times. Without gravity, energy cannot be created or organized for useful purposes. Massless energy escaping gravity cannot be expected to respond the same way it does in gravity.

Postulate O: The First State of Pure Massless Energy: Energy and Light produced inside a nuclear reaction can exist in two independent states: The first state is pure massless energy in some form of radiation that does not respond to gravity and has instantaneous velocity. This state of existence is supported by the 2nd Postulate of Relativity which states "The speed of light in a vacuum is the same for all observers, regardless of their motion relative to the source." Therefore, light is always traveling at the speed of light even if the observer is also traveling at the speed of light. This state of energy travels at infinite velocity and will extend out in proportion to the inverse square law[31] .

Postulate P: The Second State of Pure Massless Energy: This energy is the form that is converted into useful work by gravity. The second state of energy obeys the laws of physics and relativity. In this state, light is converted into photons and also obeys the Inverse Square Law.

Postulate Q: Energy traverses zero-gravity and possibly very weak gravity fields in empty space at infinite speed.

Postulate R: Gravity is the producer of wave-form energy. Gravity must necessarily transform energy into detectable forms of energy . The state of massless, pure energy itself does not exist in or possess wave forms. An as yet undefined phenomena causes gravity to react with energy producing detectable energy.

Postulate S: Independent out of phase gravity fields may cause light interference between objects in deep space and may provide additional information using the principle of "superposition" which states "When two of more waves of the same nature travel past a given point at the same time, the amplitude at the point is the sum of the instantaneous amplitudes of the individual waves."

Postulate T: Gravity affects all forms of energy in ways that can bend it which means it must have some form of mass and this mass is then burned off or dissipated as it refracts and disperses while traveling in gravity.

Postulate U: The Principle of Duality of Energy and Light: Energy and light created within a nuclear reaction of a star will respond in two distinctive ways: (1) Massless energy will accelerate outward from the star at infinite velocity escaping the stars expanding gravity field and continue to radiate outward at infinite speeds until its potential becomes ineffective due to the inverse square law. (2)The duality of energy in the presence of gravity will also produce visible light which will travel along with gravity waves at the speed of light outward from the star.

Postulate V: The interaction between energy and gravity will cease when the ratio of forces between them falls below a point of effectiveness. This ratio is stated as gravity to energy (g/e). Its values have not yet been determined.

References List

Item	Chap	Source	Subject / Application / Notes
1	intro	https://www.google.com/search?q=science+of+naturalism	In philosophy, **naturalism** is the "idea or belief that only natural laws and forces operate in the world." Adherents of naturalism assert that natural laws are the rules that govern the structure and behavior of the natural universe, that the changing universe at every stage is a product of these laws. Wikipedia
2	intro	https://biologos.org/common-questions/how-can-evolution-account-for-the-complexity-of-life-on-earth-today	A complex biological structure with many interacting parts might appear, at first glance, as if it were originally created in its present form with all its interlocking components fully formed and intact. It doesn't seem possible that they developed step by step via biological evolution. In Darwin's Black Box, Michael Behe introduces a term that he and other proponents of Intelligent Design use for this concept: irreducible complexity.
3	intro	Bible Hub Holy Bible https://biblehub.com/matthew/7-7.htm	Matthew 7:6-8 Do not give dogs what is holy; do not throw your pearls before swine. If you do, they may trample them under their feet, and then turn and tear you to pieces. 7Ask and it will be given to you; seek and you will find; knock and the door will be opened to you. 8For everyone who asks receives; he who seeks finds; and to him who knocks, the door will be opened....
4	intro	https://www.google.com/search Dark matter	Dark matter is a form of matter thought to account for approximately 85% of the matter in the universe and about a quarter of its total energy density. Wikipedia
5	intro	https://www.discovercreation.org/blog/2014/09/30/the-big-bang-theory-creation-perspective/	The **Big Bang** Theory, attempting to describe the universe's initial catastrophic event, is very complex, but still leaves a lot of questions that have to be answered. Mathematicians and Physicists are still trying to work out how (and why) the universe came from nothing, proceeded to be everything in an infinitesimally small point and then eventually expanded to what it is today.
6	intro	Bible Hub Holy Bible Genesis (all versions)	Genesis 1:1-31 "In the beginning...."
7	1	Wikipedia: https://simple.wikipedia.org/wiki/Cosmology	**Cosmology** is the branch of astronomy that deals with the origin, structure, evolution and space-time relationships of the universe.[1][2] NASA defines cosmology as "The study of the structure and changes in the present universe".[3] Another definition of cosmology is "the study of the universe, and humanity's place in it".[4] Modern cosmology is dominated by the Big Bang theory, which brings together observational astronomy and particle physics.[5]

8	2	Lance Burton, Magician	**William Lance Burton** is an American stage magician. He performed more than 15,000 shows in Las Vegas for over 5,000,000 people. In 2010 he ended his 31-year Vegas show. Wikipedia
9	2	Consensus Science – Wikipedia	**Scientific consensus** is the collective judgment, position, and opinion of the community of scientists in a particular field of study. Consensus implies general agreement, though not necessarily unanimity. Wikipedia
10	2	Creation Science: Answers in Genesis https://answersingenesis.org/creation-science/	**Creation Science** Is Real Science Evolutionists and some old-earth creationists frequently charge that scientists who believe in a young earth don't have real degrees and don't do real scientific research that can be published in peer-reviewed secular scientific journals. God Is in the Science of Mathematics If math is an invention of many human minds, then why should all these minds agree on what is correct? The Christian philosophy of math begins with God, who numbered the days of creation as recorded in Genesis 1. The founder of the true philosophy of math is Jesus Christ. Turning Secular Science into Creation Science We don't have to abandon all the work of secular scientists; we can often co-opt their work into a biblical worldview. I could now walk into any museum and—exercising caution and discernment—imagine how some of the reconstructed scenes could have been part of Noah's world. Successful Predictions by Creation Scientists Creation researchers have been studying the biblical record and the world around us to understand the history that God's Word records. While some of their theories have been discarded as new data came to light, other predictions have been powerfully confirmed.
11	3	The Journal of Clinical Investigations Dishonesty: https://www.ncbi.nlm.nih.gov/pmc/articles/PMC4639975/	**Is there dishonesty in science?**
12	3	Psychology Today: Dishonesty: https://www.psychologytoday.com/us/blog/moral-landscapes/201807/honesty-and-truthfulness-in-science	**Studies in dishonesty in science**
13	3	**"Evolution's Achilles' Heels"** Creation Book Publishers, Powder Springs, Georgia, USA www.-bookpublishers.com	PhD. Scientist explain evolution's fatal flaws-in areas claimed to be its greatest strengths. Authors" Donald Batten, PhD, Robert Carter, PhD, David Catchpool, PhD, John Hartnett, PhD, Mark Harwood, PhD, Jim Mason, PhD, Jonathan Sarfati, PhD, Emil Silvestru, PhD, Tasman, Walker, PhD.

14	3	**Science funding** is a mess: https://www.vox.com/future-perfect/2019/1/18/18183939/science-funding-grant-lotteries-research :	By some estimates, many top researchers spend 50 percent of their time writing grants. Interdisciplinary research is less likely to get funding, meaning critical kinds of research don't get done. And scientists argue that the constant fighting for funding undermines their work, by encouraging researchers to overpromise and engage in questionable practices, overincentivizing publication in top journals, disincentivizing replications of existing work, and stifling creativity and intellectual risk-taking.
15	3	**The Standard Model of the Universe**	Ref: "Creation Magazine" Creation 40(4) 2018 p19 (Sarfati): All matter in the Universe seems to be built up of 17 named fundamental particles that interact by four fundamental forces. The particles and all forces apart from gravity have been described as the "Standard model"
16	3	**The Speed of Gravity:** https://en.wikipedia.org/wiki/Speed_of_gravity	The speed of gravitational waves in the general theory of relativity is equal to the speed of light in a vacuum, c. Within the theory of special relativity, the constant c is not exclusively about light; instead it is the highest possible speed for any interaction in nature.
17	3	The **Parallax** method of distance to stars" https://en.wikipedia.org/wiki/Parallax	Parallax is a displacement or difference in the apparent position of an object viewed along two different lines of sight, and is measured by the angle or semi-angle of inclination between those two lines. Wikipedia
18	3	A **Light-Year** is a Distance: The Oxford Dictionary	a unit of astronomical distance equivalent to the distance that light travels in one year, which is 9.4607×1012 km (nearly 6 trillion miles).
19	3	**Red Shifts:** https://www.space.com/25732-red-shift-blueshift.html	Redshift and blueshift describe how light shifts toward shorter or longer wavelengths as objects in space (such as stars or galaxies) move closer or farther away from us. ... When an object moves away from us, the light is shifted to the red end of the spectrum, as its wavelengths get longer.Mar 16, 2018
20	3	**The Speed of Light:** https://www.space.com/15830-light-speed.html	The speed of light in a vacuum is 186,282 miles per second (299,792 kilometers per second), and in theory nothing can travel faster than light. In miles per hour, light speed is, well, a lot: about 670,616,629 mph. If you could travel at the speed of light, you could go around the Earth 7.5 times in one second.
21	3	**Speed of Light in Water:** https://www.quora.com/What-is-the-speed-of-light-in-water	Light travels at approximately 300,000 kilometers per second in a vacuum, which has a refractive index of 1.0, but it slows down to 225,000 kilometers per second in water (refractive index = 1.3) and 200,000 kilometers per second in glass (refractive index of 1.5).

22	3	**Coriolis Effect**: Oxford Dictionary	an effect whereby a mass moving in a rotating system experiences a force (the Coriolis force) acting perpendicular to the direction of motion and to the axis of rotation. On the earth, the effect tends to deflect moving objects to the right in the northern hemisphere and to the left in the southern and is important in the formation of cyclonic weather systems.
23	4	Book" **"Clearing Up Confusion About the Sequence of End Time Events"** by Charles N Slayton https://www.amazon.com/Clearing-Co	The Bible is no small book! If clearing up some of the confusion about the Bible is on your bucket list, this Report is a shortcut and a jump start. You will gain clear understanding of what every inquiring mind should know concerning the proper Sequence of End Time Events. It also gathers critical Scriptures that present us with a clear picture of our future before and after the End of Time. We are all appointed to die once. Believers are resurrected into eternal life; non-believers are also resurrected to face their Creator, then face punishment. God is our Judge!
24	4	**A Derivation of the Speed of Light Postulate** by"Armin Nikkah Shirazi https://deepblue.lib.umich.edu/bitstre am/handle/2027.42/83152/A_Derivat ion_of_the_Speed_of_Light_Postulat e.pdf?sequence=1	For example, the explanation for the speed of light postulate given in this paper suggests that the speed of light postulate itself should now be stated as 'the speed of light has the same value in all inertial frames of reference in space-time independent of the motion of the source or the observer'.
25		**Bible Hub Holy Bible**: He-brews 9:27-28	Just as people are destined to die once, and after that to face judgment, so Christ was sacrificed once to take away the sins of many; and he will appear a second time, not to bear sin, but to bring salvation to those who are waiting for him.
26		**Carrier wave** - Wikipedia	In telecommunications, a carrier wave, carrier signal, or just carrier, is a waveform (usually sinusoidal) that is modulated (modified) with an information bearing signal for the purpose of conveying information. This carrier wave usually has a much higher frequency than the input signal does.
27		Bible Hub **Holy Bible** NIV John 20:29	Then Jesus told him, "Because you have seen me, you have believed; blessed are those who have not seen and yet have believed."
28		**Inertial Mass:** https://www.dictionary.com/browse/i nertial-mass	Noun Physics. The mass of a body as determined by the second law of motion from the acceleration of the body when it is subjected to a force that is not due to gravity.
29	5	**Red Shifts faster than the speed of light:** https://physics.stackexchange.com/qu estions/407596/redshift	Redshifts greater than 1 are possible if the redshift is caused by relativistic motion or by cosmological expansion. ... It is caused by the expansion of space between the time when the light is emitted and when it is received by an observer.

30	8	**Definition of Photon**: Oxford Dictionary	pho·ton/ˈfōtän/ noun: photon is a particle representing a quantum of light or other electromagnetic radiation. A photon carries energy proportional to the radiation frequency but has zero rest mass.
31	9	**The Inverse Square Law**: Wikipedia	In science, an inverse-square law is any scientific law stating that a specified physical quantity is inversely proportional to the square of the distance from the source of that physical quantity. Wikipedia
32	6	**2nd Postulate of Relativity:** http://www.astronomy.ohio-state.edu/~pogge/Ast162/Unit5/sr.html Lecture 31: Special Relativity	2nd Postulate of Relativity: The speed of light in a vacuum is the same for all observers, regardless of their motion relative to the source.
33	13	**Cosmic microwave background** - Wikipediaen.wikipedia.org › wiki › Cosmic_microwave_background	The cosmic microwave background (CMB, CMBR), in Big Bang cosmology, is electromagnetic radiation as a remnant from an early stage of the universe, also known as "relic radiation". The CMB is faint cosmic background radiation filling all space. ... This glow is strongest in the microwave region of the radio spectrum.
34	intro	**Dating creation** - Wikipediaen.wikipedia.org › wiki › Dating_creation	Among the Masoretic creation estimates or calculations for the date of creation only Archbishop Ussher's specific chronology dating the creation to 4004 BC became the most accepted and popular, mainly because this specific date was attached to the King James Bible.
35	8	**Irreducible design** of the human eye https://evolutionnews.org/2020/02/the-evolution-of-the-eye-demystified/	How did the eye evolve? Michael Behe in 2006 and Jonathan Wells in 2017 wrote about the irreducible complexity of the light-sensing cascade that makes vision possible. Yet Darwinists persist in asserting that this wondrous organ emerged, without guidance or direction, from a presumed ancestral eyespot.
36	14	**Linear vs Logarithmic** - Cool Cosmoscoolcosmos.ipac.caltech.edu › cosmic_reference › linlog	Logarithmic scales are used when the actual light intensities cover such a large range that it would be difficult to represent them on a linear scale. Our eyes respond to light in a logarithmic way and see each doubling of light intensity as equal changes.
37	3	**Salvation**: Jesus Christ is our Savior	Salvation is being saved or protected from harm or being saved or delivered from a dire situation. In religion, salvation is the saving of the soul from sin and its consequences. The academic study of salvation is called soteriology. Wikipedia
38	14	**Bible Hub Holy Bible,** Isaiah 48:13, NIV	My own hand laid the foundations of the earth, and my right hand spread out the heavens; when I summon them, they all stand up together.
39	all	**"The Mainstream of Physics"** (Text Book) by: Arthur Beiser Publisher: Addison-Wesley Publishing Company, Inc.	This great old text book was referenced for most of the basic physics of light.

www.ingramcontent.com/pod-product-compliance
Lightning Source LLC
Chambersburg PA
CBHW070512160726
48003CB00004B/1537